Sol-Gel - A Versatile and Wide Technology
http://dx.doi.org/10.5772/intechopen.1003535
Edited by Susana Devesa and Sílvia Soreto

Contributors
Bárbara Costa, Francisco Romero-Gavilán, Julio Suay, Manuel Almeida Valente, Manuel P.F. Graça, Sandra Carvalho, Sara Malek, Susana Devesa, Susana Devesa, Sílvia R. Gavinho, Sílvia Soreto, Zohra Benzarti

First published in London, United Kingdom, 2025 by IntechOpen
IntechOpen is the global imprint of INTECHOPEN LIMITED, registered in England and Wales, registration number: 11086078, 167-169 Great Portland Street, London, W1W 5PF, United Kingdom

For EU product safety concerns: IN TECH d.o.o., Prolaz Marije Krucifikse Kozulić 3, 51000 Rijeka, Croatia, info@intechopen.com or visit our website at intechopen.com.

British Library Cataloguing-in-Publication Data
A catalogue record for this book is available from the British Library

Sol-Gel - A Versatile and Wide Technology
Edited by Susana Devesa and Sílvia Soreto
p. cm.

This title is part of the Materials Science Book Series, Volume 6
Topic: Metals and Nonmetals
Series Editor: Chonghe Li
Topic Editor: Joan Josep Roa Rovira

Print ISBN 978-0-85014-984-5
Online ISBN 978-0-85014-983-8
eBook (PDF) ISBN 978-0-85014-985-2
ISSN 3049-8856

If disposing of this product, please recycle the paper responsibly.

Sol-Gel - A Versatile and Wide Technology

Edited by Susana Devesa and Sílvia Soreto

Published in London, United Kingdom

IntechOpen Book Series

Materials Science

Volume 6

Aims and Scope of the Series

Materials science has always occupied an extremely high position in the human development process. As we explore the oceans of stars, various industries have put forward more stringent requirements for the performance of materials, forcing us to pay more and more attention to the development of new materials. At the same time, the formation of a data-driven scientific paradigm is dramatically shortening the development cycle of new materials. The huge data generated by synergistically combining theories, high-throughput experiments, high-throughput computation, and artificial intelligence is greatly contributing to our ability to utilize materials science to solve real-world problems. The three topics of this book series - Metals and Nonmetals; Composite Materials; and Surface Science - will address important areas of advancement in materials science. There will be a range of interesting works published under these topics.

Meet the Series Editor

Prof. Chonghe Li received his Ph.D. from the Chinese Academy of Sciences in 1995. From 1995 to 2000, he worked as a researcher at the Shanghai Institute of Metallurgy, Chinese Academy of Sciences, where he also served as director of the research laboratory. In 2000, he was appointed Professor at the Institute of High-Performance Computing in Singapore, where he worked on computation and simulation in materials science until 2004. Since then, he has been a professor at the School of Materials Science and Engineering, Shanghai University, China, as well as the director of the Shanghai Specialty Casting Engineering and Technology Research Center. Prof. Li's research focuses on titanium alloys, titanium-aluminum single crystals, intermetallic compounds, theoretical calculations, alloy design, and special refractory materials. His broad scientific expertise is well recognized by the scientific community around the world. He is a member of the editorial board of the journal Metals. As an author, he has published more than 200 peer-reviewed papers, 2 books, and over 40 patents.

Meet the Volume Editor

Susana Devesa holds a degree in Physics and Chemistry, a Master's in Applied Physics, a Master's in Civil Engineering, and a Ph.D. in Physics Engineering. She is currently a researcher in the Mechanical Engineering Department at the University of Coimbra and a member of the interdisciplinary R&D Unit Centre for Mechanical Engineering, Materials and Processes (CEMMPRE). Her research interests are diverse and encompass several critical areas, including the green synthesis of advanced ceramic materials and metal oxides, as well as the development of flexible and sustainable functional films for applications in industries such as electronics, biomedicine, and environmental technology. Additionally, she is involved in the electrodeposition and characterization of functional coatings, which enhance the performance and durability of various substrates.

Sílvia Soreto, a researcher at the University of Aveiro(UA)/i3N, specializes in synthesizing and characterizing functional materials using sol-gel and related techniques. She earned her Bachelor's in Teaching Physics and Chemistry in 2003. She began her research during her Master's (2006) at UA and FCT-UNL, employing impedance spectroscopy to study polymeric materials' glass transition parameters. During her Ph.D. (2019) at UA with an FCT grant, she developed ferrite-based materials for electronic applications, focusing on enhancing magnetic and electric energy storage. Her research later expanded to biomedical applications, exploring ferrite-based materials in magnetic hyperthermia therapy for cancer treatment. Dr. Soreto's expertise includes solid-state physics, sol-gel synthesis, and micro- and nanostructured materials processing. She also gained industrial experience as a researcher, working on flash sintering techniques for ceramic processing.

Contents

Preface

The sol-gel method is a wet chemical process based on the production of a homogeneous sol from the precursors and its conversion into a gel, suitable for the synthesis of various nanostructures, especially oxide-based materials.

The sol-gel process allows the fabrication of advanced materials in a wide variety of forms, such as ultrafine or spherical-shaped powders, thin film coatings, fibers, porous or dense materials, and extremely porous aerogel materials.

This versatility, combined with other advantages, including low-cost, flexible synthesis routes and control over the properties of the materials, shows the tremendous potential of the sol-gel method and justifies its application in many areas, such as optics, electronics, ionics, mechanics, energy, environment, biology, medicine, pharmaceutical, functional smart coatings, among others.

Recently, sustainable approaches have gained increasing attention, leading to the development of green sol-gel synthesis routes. These innovative routes replace hazardous reagents with environmentally friendly alternatives, such as plant extracts, biopolymers, and water-based solvents. The use of natural compounds as polymerizing and reducing agents allows for a more sustainable process, reducing toxicity and environmental impact while maintaining excellent control over the material properties. Such green strategies enhance the potential of sol-gel technology for applications in biomedicine, catalysis, and eco-friendly functional materials.

This book aims to provide the reader with a comprehensive overview of the current state-of-the-art development of advanced materials using sol-gel technology, highlighting its broad applicability.

Chapter 1, "Comparative Overview of Metal Oxide Nanoparticle Synthesis Methods: Conventional Sol-Gel Versus Green Approaches", explores both conventional and green sol-gel synthesis approaches for metal oxide nanoparticles, focusing on CuO, Cu_2O, ZnO, TiO_2, MgO, and NiO, with an emphasis on the types of plants used and the procedures employed to obtain the extracts.

Chapter 2, "Application of Silica Sol-Gel Coatings for the Bioactivation of Metallic Implants", describes the main advances in the development and characterization of silica coatings applied on medical implants by the sol-gel method to enhance their performance.

Chapter 3, "Proteic Sol-Gel Route: An Eco-friendly Method to Synthesize Magnetic Nanoparticles", provides a comprehensive review of the proteic Sol-Gel process and presents a case study on the preparation of an innovative magnetic composite consisting of gadolinium ferrite and zinc ferrite, both synthesized via this eco-friendly route, using powdered coconut water as the polymeric agent.

Chapter 4, "Synthesis of Fluorine-Doped Tin Oxide Powders by Sol-Gel Combustion Method", presents the successful sol-gel combustion synthesis of fluorine-doped tin oxide nanoparticles, with an average size of 20 nm and equiaxed morphology, through the optimization of the synthesis parameters.

Chapter 5, "Sol-Gel Synthesis of Hydroxyapatite: Applications, Methods, and a Case Study", discusses different sol-gel synthesis routes of hydroxyapatite, emphasizing the influence of precursor materials and solvent systems on the phase purity and morphology of the resulting materials. Additionally, a detailed case study on the synthesis and dielectric properties of hydroxyapatite is presented.

Susana Devesa
ARISE, Department of Mechanical Engineering,
University of Coimbra, CEMMPRE,
Coimbra, Portugal

Sílvia Soreto
i3N and Department of Physics,
University of Aveiro,
Aveiro, Portugal

Chapter 1

Comparative Overview of Metal Oxide Nanoparticle Synthesis Methods: Conventional Sol-Gel Versus Green Approaches

Susana Devesa

Abstract

Nanoparticles are fundamental in nanotechnology, with their applications expanding rapidly due to their unique optical, chemical, electrical, and mechanical properties. Metal oxide nanoparticles are widely studied due to the abundance of their natural precursors, environmental friendliness, affordability, and favorable electrical and optical properties. In some cases, they also exhibit chemical stability. Recent advances in synthesis methods, particularly the sol-gel technique, have enabled precise control over nanoparticle size and shape, offering benefits such as simplicity, low cost, and lower processing temperatures. More recently, eco-friendly biosynthesis of nanoparticles has garnered increasing attention due to its numerous advantages over traditional chemical and physical methods, which often involve hazardous and expensive chemicals. Green synthesis using biological entities like plants, algae, bacteria, fungi, and viruses offers several benefits. The use of plant extracts, in particular, is advantageous because they are readily available, safe to handle, and capable of producing large quantities of nanoparticles. They can act as both reducing and stabilizing agents during synthesis. This mini-review explores both conventional and green sol-gel synthesis approaches for metal oxide nanoparticles, focusing on CuO, Cu_2O, ZnO, TiO_2, MgO, and NiO, with an emphasis on the types of plants used and the procedures employed to obtain the extracts.

Keywords: nanotechnology, nanoparticles, oxide nanoparticles, sol-gel technology, green synthesis

1. Introduction

According to the Oxford English Dictionary, the linguistic form "nano" originates from the classical Latin *nanus* or its ancient Greek etymon *nanos*, meaning "dwarf" [1, 2].

In 1947, during the 14th International Union of Pure and Applied Chemistry (IUPAC) conference, the prefix nano was officially designated to represent one billionth (10^{-9}) of a unit [3]. In 1958, the prefixes nano, giga, tera, and pico were adopted into the newly established International System of Units [1].

Nano has become a popular label in modern science, with many terms incorporating the prefix now appearing in dictionaries. These include nanometer, nanoscale, nanoscience, nanotechnology, nanostructure, nanotube, nanowire, and nanorobot, among others [2]. In these contexts, the prefix nano is used to denote "very small" entities or processes, typically on the nanometer scale [3].

In scientific literature, the prefix "nano" has become a widely adopted label across various fields to describe small entities and processes. This includes terms such as nanoscience, nanotechnology, nanorobots, nanomagnets, nanoelectronics, and nanoencapsulation, among others [1, 3]. Nano is used to signify "very small" entities or processes, often on the nanometer scale. The popularity of this prefix has led to its inclusion in many dictionaries, with terms like nanometer, nanoscale, nanostructure, nanotube, nanowire, and nanorobot becoming commonplace in modern science.

The primary concept of nanotechnology was put forward by Richard Feynman in 1959 while presenting his famous talk at the annual meeting of the American Physical Society [4]. Though he never explicitly mentioned nanotechnology, Feynman suggested that it will eventually be possible to precisely manipulate atoms and molecules [5].

In 1974, Norio Taniguchi introduced the term *nanotechnology* at an engineering conference in Tokyo while hypothesizing the idea of precision manufacturing at the scale of nanometers [1, 4, 6].

Later, nanotechnology as a discipline was firmly established through the contributions of Richard Smalley, Eric Drexler, and Chad Mirkin [4, 7].

Nowadays, nanotechnology and nanoscience are two specialized terms widely used to describe a broad range of research activities and innovations worldwide [8].

In 1994, the Royal Society/Royal Academy of Engineering Working Group provided clear definitions to differentiate between nanoscience and nanotechnology. Nanoscience was defined as the study of structures and molecules at the atomic, molecular, and macromolecular scales, where properties differ significantly from those observed at larger scales. In contrast, nanotechnology refers to the application of nanoscience in practical contexts, such as the development of various devices or systems, by precisely controlling shape and size at the nanometer scale [9].

One of the fundamental concepts in nanotechnology is nanoparticles [8]. Currently, the application of nanoparticles is well-established and continues to expand due to their distinctive optical, chemical, electrical, electronic, and mechanical properties. These unique characteristics are largely a result of their high surface area-to-volume ratio, which gives them properties that differ significantly from those of individual atoms or molecules and from the bulk material [6].

Nanoparticles are routinely defined as single particles with a diameter between about 1 and 100 nm that show properties that are not found in bulk samples of the same material [10, 11].

This upper limit is established based on the concept that, at these sizes, the physical properties of nanoparticles begin to resemble those of bulk materials. However, some sources argue that this size threshold can vary depending on the material and thus should not be the only criterion for classification. A more comprehensive definition extends this upper size limit to 1 μm, with the submicron range still being classified as nano [2].

However, the International Organization for Standardization (ISO) defines nanoparticles as nano-objects with all external dimensions in the nanoscale, which is from 1 nm to 100 nm, where the lengths of the longest and the shortest axes of the nano-object do not differ significantly. If the dimensions differ significantly, terms such as nanofibers, for two external dimensions in the nanoscale, and nanoplates, for three external dimensions in the nanoscale, should be applied instead of the term nanoparticle [3, 12].

Figure 1 shows, schematically, this hierarchy of terms related to nano-objects.

Nanoparticles are broadly divided into various categories depending on their morphology and chemical properties [13]. Regarding the morphology, they can be categorized according to their shape, including flower, spherical, cubic, rod, triangular, pentahedral, octahedral, star, etc. [8, 14]. Based on their composition, three major classes can be considered: organic, carbon-based, and inorganic nanoparticles [3].

The organic nanoparticles, which are typically non-toxic and biodegradable, are composed of proteins, carbohydrates, lipids, polymers, or any other organic compounds [3, 15]. Some prominent examples of this class are dendrimers, liposomes, and micelles [3, 15, 16].

The carbon-based class comprises nanoparticles that are made solely from carbon atoms. Well-known examples of this class include fullerenes, carbon black nanoparticles, and carbon quantum dots [3, 15].

This inorganic class comprises nanoparticles that do not contain carbon or organic materials in their composition. Typical examples of this class include metal, ceramic, and semiconductor nanoparticles [3].

Metal nanoparticles, as the name implies, are composed purely of metal precursors and can be monometallic, bimetallic, or polymetallic, while metal oxide nanoparticles consist of a metal precursor combined with oxygen [3, 9].

Metal and metal oxide nanoparticles exhibit a broad range of physicochemical properties, surpassing their bulk counterparts in terms of chemical, electrical, optical,

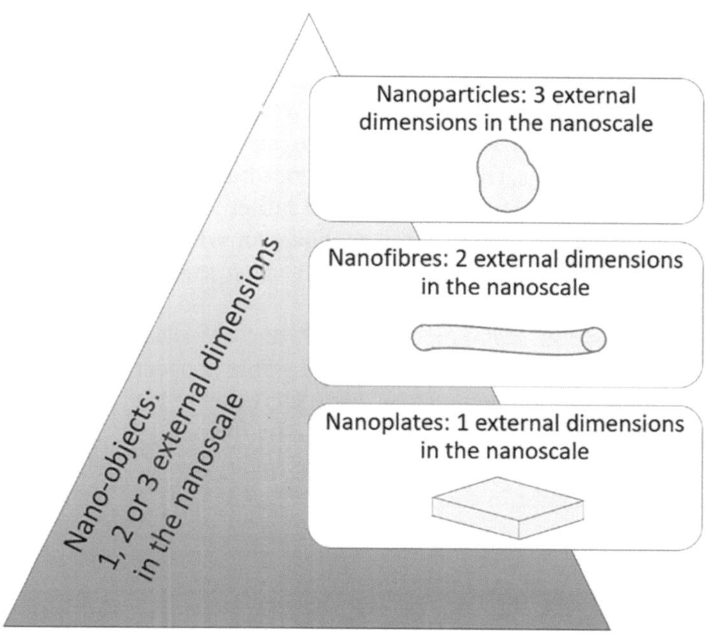

Figure 1.
Diagram illustrating the types of nano-object materials based on their dimensions, according to the International Organization for Standardization [12].

thermal, mechanical, electromagnetic, and surface characteristics. They also provide large surface areas, controlled size and shape, and easy surface modification. Due to these advantages, they find applications in fields such as biomedicine, catalysis, environmental remediation, energy harvesting, molecular sensing, and more [9, 17].

Metal oxide nanoparticles are an interesting class and distinct from metal nanoparticles, being considered as one of the most emerging nanomaterials in many areas [18].

Metal oxide nanoparticles are extensively studied due to the abundance of their precursors in nature, their environmental friendliness, affordability, favorable electrical and optical properties, and, in certain instances, their chemical stability [19, 20].

Furthermore, these materials exhibit remarkable characteristics such as resistance to mechanical stress, excellent optical transparency, high carrier mobility, a wide band gap, and a high dielectric constant, along with superconductivity and other properties. As a result, metal oxides are regarded as some of the most intriguing functional materials, making them highly valuable in a wide range of technological applications [19, 21].

Metal oxides are ionic compounds, composed of positively charged metal ions and negatively charged oxygen ions. The strong electrostatic attraction between these oppositely charged ions leads to the formation of solid and robust ionic bonds [21, 22]. They can exhibit metallic, semiconductor, or insulator characteristics, with the possibility of engineering their electronic properties by modifying their morphology, stoichiometry, and through doping [23].

Semiconductor metal oxides can be categorized into two types: n-type and p-type. The n-type semiconductor metal oxides have oxygen-vacancy donors responsible for n-type conductivity, and the majority of charge carriers are electrons. In contrast, p-type semiconductors are metal-ion-deficient, and the majority of charge carriers are holes [24].

Nevertheless, the electronic structure range of these materials is extensive, being divided into two main categories, transition and non-transition-metal oxides, where the latter englobes the pre- and post-transition-metal oxides [25].

In transition-metal oxides, while the s-shells of positive metal ions are always fully occupied by electrons, their d-shells may remain partially unfilled [26].

In transition-metal oxides, the energy difference between a cation d^n configuration and either a d^{n+1} or d^{n-1} configuration is often rather small, which allows a fast transformation between the different forms, however with unstable structures. This feature makes them more sensitive than pre-transition-metal oxides to the environment [19, 25].

Pre- and post-transition-metal oxides with d^0 and d^{10} electronic configurations, respectively, are characterized as materials with stable properties. Pre-transition-metal oxides are generally expected to be quite inert due to their large band gaps, which make it difficult for electrons and holes to be formed [19, 25].

The various types of metal oxides include CuO, Cu_2O, ZnO, TiO_2, MgO, and NiO, which will be addressed later in this chapter.

2. Synthesis of nanoparticles

In recent decades, significant efforts have been devoted to developing reliable synthetic methods for nanosized materials, allowing precise control over their size, shape, and other unique properties [18].

There are different methods that can be employed for the synthesis of nanoparticles, which are broadly divided into two main classes: (i) the bottom-up approach and (ii) the top-down approach [9], as schematically represented in **Figure 2**.

In the top-down synthesis, a destructive approach is used, where the macroscopic material is reduced by physical or chemical means into smaller units, which are then converted into suitable nanoparticles. Examples of this method include thermal decomposition, mechanical grinding/milling, physical vapor deposition, chemical etching, laser ablation, electro-explosion, and lithography [13, 16, 17].

The top-down approach is considered simple; however, it usually requires expensive equipment, being its main disadvantage the final imperfection of the surface structure, as defects in the surface structure can significantly affect the final physical properties and surface chemistry of the nanoparticles due to their high aspect ratio [9, 17].

The bottom-up synthesis, also known as the constructive method, is the reverse of the top-down approach. This method is based on processes of self-assembly of individual atoms into larger clusters and their further aggregation into final nanoparticles. The bottom-up approach includes methods such as chemical vapor deposition, sol-gel, spinning, pyrolysis, and biological synthesis [11, 13, 16, 17]. A distinct advantage of this approach is the increased possibility of obtaining metallic nanoparticles with fewer defects and a more homogeneous chemical composition; however, it requires compatible surfaces and molecules [9, 17].

2.1 Conventional sol-gel method

The sol-gel technique is a wet-chemical, bottom-up process that is widely utilized as both a traditional and commercial method for producing various nanostructures,

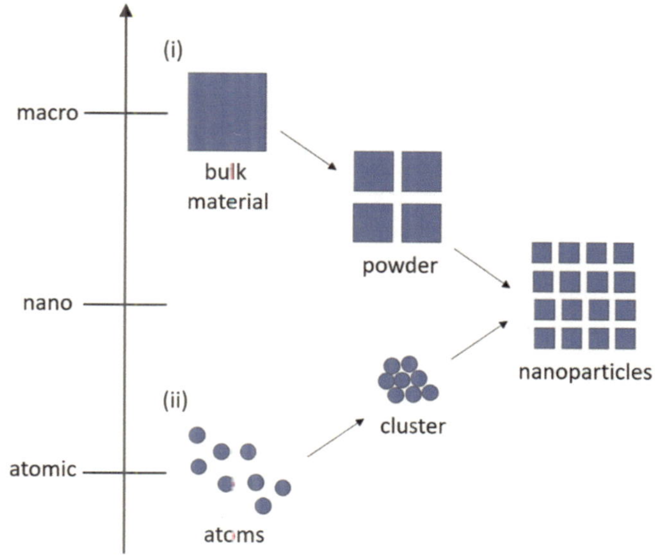

Figure 2.
Schematic representation of the two main classes of nanoparticle synthesis: (i) the top-down approach and (ii) the bottom-up approach, based on [9, 27].

especially metal oxide nanoparticles. It offers excellent control over particle size, high homogeneity, and purity at low temperatures [28, 29].

Sol-gel, as the name suggests, involves two concepts: sol and gel. A sol is a colloidal suspension or dispersion of small particles in a liquid medium, making it a subclass of colloidal solutions. A gel, on the other hand, is a porous, three-dimensional, continuous solid network that surrounds a continuous liquid phase. The sol-gel process involves gradually converting the sol into a gel-like continuous network (gelling or viscous state) before reaching the final stage [30, 31].

Sols and gels have sparked scientific interest for a long time. The oldest sols prepared in a laboratory were synthesized with gold by Faraday in 1853. It was not until 1861 that Graham founded colloidal science.

While some silica gels occur naturally, the ability to synthesize them was not developed until the nineteenth century. The first synthetic silica gels were created by Ebelmen in 1846, and Cossa followed with the synthesis of alumina gels in 1870 [32].

Historical accounts commonly credit Ebelman with the first documented sol-gel reaction. Ebelman, in 1846, accidentally left tetraethyl orthosilicate unattended, and upon returning several months later, he discovered a substance that resembled glass. This material had a density exceeding 1.8 and was hard enough to scratch glass. At that time, the chemical understanding was insufficient to accurately determine the formula of this "silica," which was indeed a gel but with a high degree of condensation [33, 34].

In 1939, Geffcken and Berger patented a technique for creating thin films from alkoxides, extending beyond silica to include materials like alumina and titania. After World War II, Glaswerke, under the guidance of Dislich, adapted these foundational techniques to develop an industrial process for applying functional coatings to flat glass. This innovation drew the attention of the glass industry and sparked a surge of scientific research into glass coatings, effectively marking the birth of the sol-gel field [33, 34].

Since 1971, it has been established that the sol-gel process can be used to synthesize a variety of multicomponent oxides using alkoxides of different elements [35]. However, the field saw a major milestone in 1981 with the inaugural International Conference on Glasses and Glass Ceramics Obtained from Gels, held in Padova, Italy. This event is considered a pivotal moment in the advancement of sol-gel science. Additionally, the Journal of Sol-Gel Science and Technology, which has been published regularly since 1993, being the founding editor S. Sakka, plays a crucial role in disseminating a broad spectrum of research and developments in this area [33].

Indeed, according to Sakka, considered one of the pioneers in the development of sol-gel technology, sol-gel processes are considered a form of nanotechnology because the resulting gels often contain nanoparticles or form nanocomposites. In this context, sol-gel technology plays a crucial role in advancing modern nanotechnology, particularly in the preparation of new materials [33, 36].

The sol-gel technique is highly popular among researchers for producing nanomaterials such as nanoparticles, nanofibers, nanofilms, aerogels, and nanoporous materials, often without the need for melting or sintering [31].

Considering the metal oxide nanoparticles, in the sol-gel process, they are formed through a series of irreversible chemical reactions, as shown in the schematic representation in **Figure 3**.

The precursors are dispersed in a host solvent by shaking, stirring, or sonication, leading to the formation of a mixture in the hydrolysis step, known as the sol [15, 17, 21, 29], which is a stable suspension of colloidal solid particles.

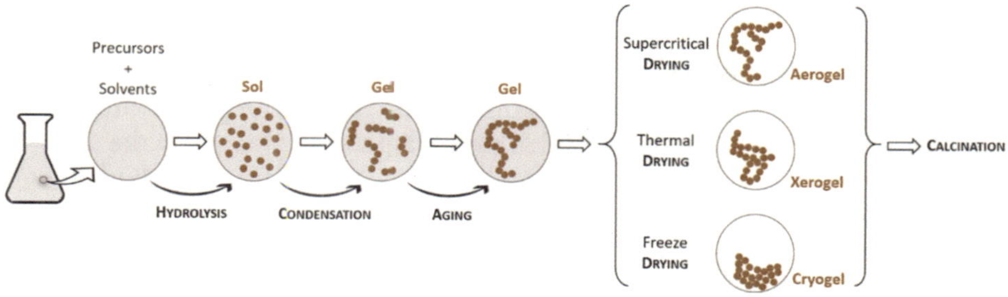

Figure 3.
Schematic representation of the different stages of the sol-gel process.

The hydrolysis process of precursors such as metal alkoxides in the host solvent can be water or organic solvents (e.g., alcohols). For the synthesis of metal oxide nanoparticles, oxygen is necessary for the formation of the metal oxide, which is supplied by water or organic solvent. In this step, a nucleophilic attack occurs on the metal alkoxide by the oxygen. This leads to the simultaneous removal of the alkoxide functional group and the formation of the alcohol functional group (M − OH) [37, 38].

In addition to water and alcohol, an acid or a base also facilitates the hydrolysis of the precursors [37]. Typically, the hydrolysis process tends to slow down in acidic conditions and accelerate in basic conditions [38].

The general chemical reaction for the hydrolysis process can be written as [37, 38]:

$$\text{M-OR} + H_2O \rightarrow \text{MOH} + \text{ROH} \tag{1}$$

where M = metal and R = alkyl group (C_nH_{2n+1}).

The formed sol then transitions into a semi-solid phase through the condensation process, forming a gel [15, 17, 21, 29].

This step involves condensation of adjacent molecules, where water/alcohol are eliminated and metal oxide linkages are formed, and polymeric networks grow to colloidal dimensions. Condensation occurs via two processes: olation, where a hydroxyl (–OH–) bridge is formed between two metal centers (metal–hydroxy-metal bonds), and oxolation, a process in which an oxo (–O–) bridge is formed between two metal centers (metal–oxo-metal bonds) [38].

The general chemical reaction for the condensation process can be represented as follows [37, 38]:

$$\text{M-OH} + \text{XO-M} \rightarrow \text{M-O-M} + \text{XOH} \tag{2}$$

where M = metal and X = H or alkyl group (C_nH_{2n+1}).

Subsequently, the aging process leads to ongoing changes in the structure and properties of the gel. During this process, polycondensation continues within the localized solution, accompanied by the reprecipitation of the gel network. As a result, porosity decreases and the thickness between colloidal particles increases [37].

The phase separation that occurs between the solvent and the gel can be implemented by sedimentation, filtration, or centrifugation. The remaining moisture is removed through drying. Depending on the desired characteristics and applications of the resulting nanoparticles, appropriate drying is essential [15, 28].

There are different drying processes, each affecting the structure of the gel network in distinct ways, as shown in **Figure 3**: supercritical drying, thermal drying, and freeze-drying [37, 38].

In supercritical drying, aerogels are formed, which have high pore volume and surface area, with the original gel network remaining almost intact. On the other hand, heating the porous gel at high temperature causes densification, eliminating the pores from the gel to get the xerogel. Also, the drying can be done by freezing the solvents to form the cryogel [37].

Finally, thermal treatment, or calcination, is used to remove residual materials and water from the sample. The calcination temperature is a critical factor in determining the pore size and density of the final material [37].

The sol-gel method offers several advantages, including simplicity, low cost, high homogeneity, and the production of nanoparticles with high purity. Additionally, one of its key benefits over conventional methods is the lower processing temperature. Due to its unique properties, this method is capable of producing high-quality nanoparticles of uniform size on an industrial scale [17, 29].

2.2 Green approaches

In 1998, Paul Anastas and John Warner outlined twelve principles of "green" chemistry in their book *Green Chemistry: Theory and Practice*. They advise scientists, industry professionals, and policymakers to focus on minimizing or eliminating the use of hazardous materials and processes. These principles have played a crucial role in the development and advancement of a new philosophical approach due to their relevance, practicality, and specificity [8, 39].

The twelve green chemistry principles are as follows [40]:

1. "It is better to prevent waste than to treat or clean up waste after it has been created."

2. "Synthetic methods should be designed to maximize the incorporation of all materials used in the process into the final product."

3. "Wherever practicable, synthetic methods should be designed to use and generate substances that possess little or no toxicity to human health and the environment."

4. "Chemical products should be designed to affect their desired function while minimizing their toxicity."

5. "The use of auxiliary substances (*e.g.*, solvents, separation agents, *etc.*) should be made unnecessary wherever possible and innocuous when used."

6. "Energy requirements of chemical processes should be recognized for their environmental and economic impacts and should be minimized. If possible, synthetic methods should be conducted at ambient temperature and pressure."

7. "A raw material or feedstock should be renewable rather than depleting whenever technically and economically practicable."

8. "Unnecessary derivatization (use of blocking groups, protection/deprotection, temporary modification of physical/chemical processes) should be minimized or avoided if possible, because such steps require additional reagents and can generate waste."

9. "Catalytic reagents (as selective as possible) are superior to stoichiometric reagents."

10. "Chemical products should be designed so that at the end of their function they break down into innocuous degradation products and do not persist in the environment."

11. "Analytical methodologies need to be further developed to allow for real-time, in-process monitoring and control prior to the formation of hazardous substances."

12. "Substances and the form of a substance used in chemical processes should be chosen to minimize the potential for chemical accidents, including releases, explosions, and fires."

Developing rapid, safe, non-toxic, and environmentally friendly methods for producing metal oxide nanoparticles is consistent with the principles of green chemistry [41]. Biosynthesis of nanoparticles using eco-friendly techniques has gained increasing attention due to its numerous advantages over traditional chemical and physical methods, which often involve hazardous, expensive chemicals and/or complex equipment [42].

Green synthesis of nanoparticles through biological entities, such as plants, algae, bacteria, fungi, and viruses, offers several benefits. The use of plant extracts has some advantages, as they are readily available, safe to handle, and capable of producing large quantities of nanoparticles. They can also serve as both reducing and stabilizing agents in the synthesis process [42]. Key naturally occurring organic compounds involved in this synthesis include terpenoids, flavonoids, alkaloids, and phenolic compounds [8, 42].

To synthesize nanoparticles using extracts from plants, the extract is mixed with solutions of the metal ions precursor under various reaction conditions. Key factors that are recognized as important in controlling the rate, yield, and stability of nanoparticle formation are the metal salt concentration, temperature, and pH, but also the types of phytochemicals present in the plant extract employed [8, 43].

The process of bio-producing metal oxide nanoparticles using plant materials can be divided into three stages. First, during the activation stage, metal ions are reduced from a mono- or divalent state to a zero-valent state. This reduction leads to nucleation. In the growth stage, small nanoparticles aggregate and evolve into larger particles with diverse shapes such as pentagonal, hexagonal, spherical, cubic, or triangular. Plant extracts also function as stabilizing and capping agents, ensuring stable nanoparticle morphology. Finally, in the termination stage, nanoparticle aggregates form various irregularly shaped nanoparticles [8, 44].

The green synthesis approach, irrespective of the biological entity used, is favored for its cost-effectiveness, reduced toxicity, environmental friendliness, and improved applicability compared to conventional chemical methods [44], namely in what concerns biocompatibility and crucial integration with various medical applications [45].

An analysis of papers published in SCOPUS over the last ten years, as shown in **Figure 4(a)**, reveals a growing interest in the synthesis of metal oxide nanoparticles. The popularity of the conventional sol-gel method among various preparation techniques has, however, stabilized in the last couple of years.

Figure 4(b) highlights the increasing number of reports on the synthesis of metal oxide nanoparticles using green approaches, showing also that publications involving plant extracts, in particular, have made a significant contribution. However, it is important to note that not all of these green approaches are necessarily based on sol-gel technology.

The number of published articles was retrieved from the SCOPUS database using the fields "Article title, Abstract, Keywords," for the period from 2014 to 2023, with the combination of keywords identified in the respective plots.

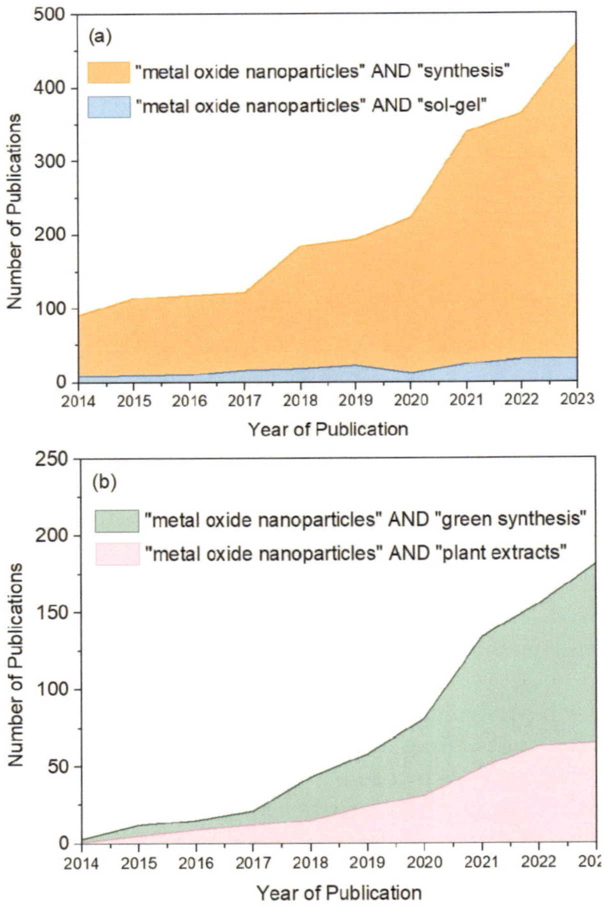

Figure 4.
Number of published articles retrieved using the specified keywords (SCOPUS database): (a) "metal oxide nanoparticles" AND "synthesis" and "metal oxide nanoparticles" AND "sol–gel"; (b) "metal oxide nanoparticles" AND "green synthesis" and "metal oxide nanoparticles" AND "plant extracts."

3. Conventional and green sol-gel methods for metal oxide nanoparticle synthesis

This mini-review focuses on metal oxide nanoparticles of significant interest, presenting both conventional and green sol-gel synthesis methods and the starting materials used. The discussion of the green approaches will be centered on methods using plant extracts, with particular emphasis on the type of extract used and the procedure employed to obtain it.

Additionally, features such as particle size and reported outcomes related to the targeted applications will be presented when available.

Figure 5 shows a diagram that visually represents the different metal oxide nanoparticles covered, CuO, Cu_2O, ZnO, TiO_2, MgO, and NiO, and the plant extracts used for their green synthesis.

3.1 Copper(II) oxide, CuO, nanoparticles

Copper(II) oxide is a p-type semiconductor with a band gap ranging from 1.2 to 2.1 eV [46, 47]. CuO possesses several distinctive properties, such as antimicrobial activity, efficient light absorption, and low cost, making it a promising material among metal oxides [46].

CuO nanoparticles find applications in microwave irradiation, gas sensing, redox catalysis, and various oxidation processes. They are also utilized in photoconductive and photothermal applications, demonstrating effectiveness in adsorption processes

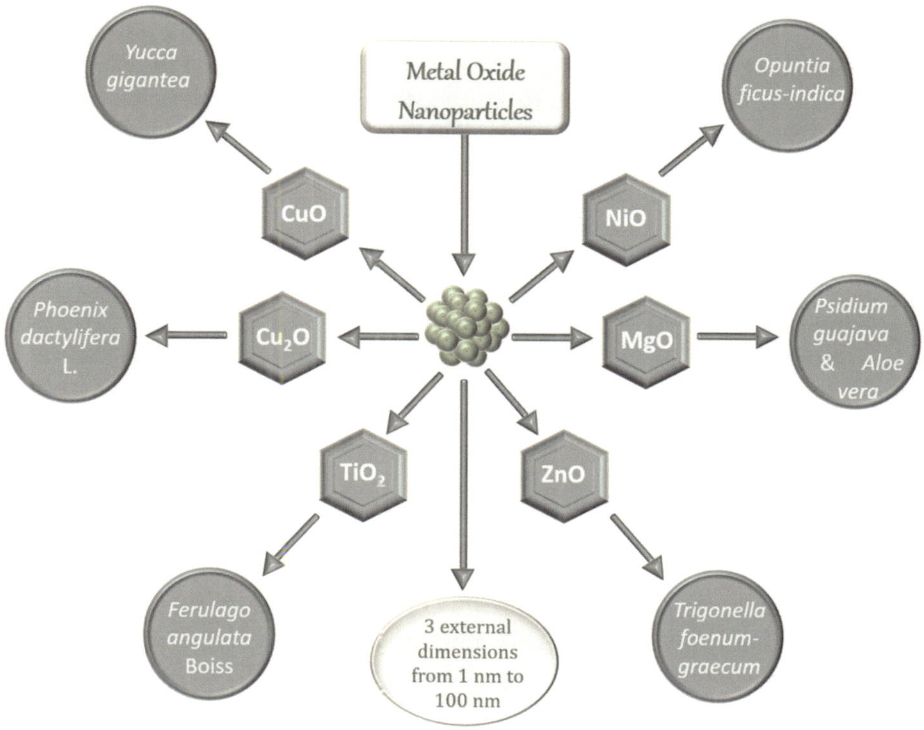

Figure 5.
Schematic representation of the metal oxide nanoparticles discussed and the plants used in their biosynthesis.

[47, 48]. The small size, high surface area, natural abundance of raw materials, cost-effective production, and non-toxic nature of CuO nanoparticles further enhance their appeal [49].

Wang et al. [50] reported the synthesis of CuO nanoparticles using the conventional sol-gel method. The starting materials used were copper acetate and isopropyl alcohol, with ethanolamine added as a stabilizer. After drying in an oven, the resulting mixture was annealed at 500°C for 2 hours.

Structural analyses performed using X-ray diffraction, Raman spectroscopy, and X-ray photoelectron spectroscopy showed that the obtained sample is crystalline and has the monoclinic structure of CuO. Meanwhile, morphological characterization, conducted using scanning electron microscopy, revealed particles with an average diameter of approximately 100 nm.

According to the authors, the CuO nanoparticle sensor developed exhibited excellent gas-sensing behavior toward acetone, methanol, and ethanol, suggesting that CuO nanoparticles may be a promising candidate for sensing low concentrations of reducing volatile organic gases using a simple and scalable production process.

Alotaibi et al. [45] employed a green sol-gel method to synthesize copper oxide nanoparticles. In their study, the starting materials used were copper(II) chloride dihydrate, sodium hydroxide, and an extract from *Yucca gigantea*. The obtained precipitate was dried at 100°C and heat-treated at 500°C for 4 hours.

The authors reported that the *Yucca gigantea* extract was used in a scalable method to produce CuO nanoparticles in a manner that is safe, cost-effective, sustainable, and practical.

The *Yucca*, a genus of perennial shrubs and trees belonging to the Asparagaceae family, is predominantly found in arid regions of the Southwestern United States and Mexico and includes approximately 40 to 50 species with medicinal properties [51].

For this study [45], *Yucca gigantea* leaves were dried and ground into a fine powder. This powder was subjected to a three-stage extraction process using a solvent mixture of 70% methanol and 30% water. The resulting aqueous extract was then evaporated under reduced pressure, yielding a residue of 24.66%, which was stored at −20°C for subsequent analysis.

The presence of CuO nanoparticles was confirmed by an infrared spectral band at 515 cm^{-1}, indicating the Cu–O metal–oxygen bonds and showing the *Yucca gigantea* potential effect in the reduction of metal precursors for the fabrication of metal oxide nanoparticles.

The scanning and transmission electron microscopy analyses revealed that the nanoparticles ranged in size from 30 to 50 nm, with the in vivo studies demonstrating that these CuO nanoparticles significantly reduced bacterial load and improved wound healing rates.

3.2 Copper(I) oxide, Cu_2O, nanoparticles

Copper(I) oxide, or cuprous oxide, is also a p-type semiconductor with a band gap ranging from 1.2 to 2.1 eV [46]. Cu_2O is considered one of the most important materials used in contemporary technologies and is widely utilized in fields such as superconductors, catalysis, fuel cells, photovoltaics, optoelectronics, magnetic materials, pigments, sensors, and ceramics. Additionally, Cu_2O nanoparticles are known to act as antimicrobial agents due to their enhanced antimicrobial activity against various pathogenic microorganisms [52].

Zayyoun et al. [53] synthesized Cu_2O nanoparticles using the sol-gel method in a glycolic medium. Copper(II) acetate monohydrate and ethylene glycol were used as starting materials. Subsequently, different amounts of sodium hydroxide were added to the solution to vary the pH.

At lower pH values, below 7, an orange solution formed, which the authors identified as indicative of Cu_2O particle synthesis. The resulting precipitates were filtered, washed twice with distilled water and once with absolute ethanol to remove impurities, and then dried at 50°C.

X-ray diffraction characterization of the obtained samples confirmed the presence of pure Cu_2O under acidic conditions (pH = 6.1). The morphology of this powder, observed via transmission electron microscopy, revealed very small cubic particles with an average diameter of 3.24 nm, which aggregated into spherical shapes due to the effect of ethylene glycol.

The Cu_2O nanoparticles exhibited notable optical properties, with a band gap value of 2.07 eV determined from Tauc's plot (higher as compared to the bulk value), suggesting their potential for applications in photoelectronic devices and solar energy harvesting.

Berra et al. [54] synthesized Cu_2O nanoparticles using a green synthesis route involving phenolic compounds from leaf extract to reduce copper sulfate. The starting materials used were copper(II) sulfate pentahydrate and an extract from the leaves of *Phoenix dactylifera* L. The obtained precipitate was washed several times with absolute ethanol and distilled water, then dried overnight at 80°C and heat-treated at 400°C for 2 hours.

Phoenix dactylifera L., widely known as the date palm, is among the oldest cultivated fruit plants, thriving in arid regions like the Arabian Peninsula, North Africa, and the Middle East [55]. Dates are highly nutritious, providing abundant carbohydrates, dietary fiber, and essential vitamins and minerals. The pits of dates also offer substantial amounts of minerals, lipids, and proteins. Moreover, dates have been utilized in traditional medicine for their therapeutic properties. Phytochemical studies have identified a variety of beneficial compounds in dates, including anthocyanins, phenolics, sterols, carotenoids, procyanidins, and flavonoids [56].

In the study mentioned above [54], fresh leaves were washed and dried at room temperature for 15 days. The plant material was extracted using the maceration method with 70% ethanol solution, maintaining a ratio of plant material mass to solvent volume of 1:6. The mixtures were kept in a tightly sealed vessel at room temperature for 24 hours before use. The authors determined that the volume ratio of plant extract to copper sulfate required to produce Cu_2O nanoparticles is between 1:50 and 1:70. XRD characterization confirmed the formation of cubic Cu_2O.

Scanning electron microscopy analysis showed that the synthesized particles are roughly spherical, with shapes varying from spherical nanoparticles to small agglomerations as the concentration of plant extract increased. While the particles appear to be nanosized, the authors did not specify the average grain size.

3.3 Zinc oxide, ZnO, nanoparticles

ZnO is an n-type semiconductor oxide with a broad direct band gap of 3.37 eV [57, 58]. It is an excellent semiconductor oxide, possessing remarkable electrical, mechanical, and optical properties. Moreover, ZnO not only exhibits antifouling and antibacterial properties but also demonstrates good photocatalytic activity [57].

Due to its remarkable characteristics, such as high thermal conductivity, high refractive index, strong binding energy, UV shielding, and antibacterial effects, zinc oxide finds applications across a wide range of industries, including medicine, cosmetics, solar energy, rubber, concrete, and food products [58].

The unique characteristics of ZnO nanoparticles make them a promising candidate for various applications in biomedicine, particularly for their low toxicity, biocompatibility, biodegradability, wound healing, antibacterial, antineoplastic, and antigenic properties [58, 59].

In the study conducted by Chung et al. [59], the synthesis of ZnO nanoparticles using a sol-gel method was optimized through the application of the Taguchi design.

Zinc acetate dihydrate and oxalic acid mixtures were prepared in ethanol, and the resulting gel was dried at 60°C overnight. The gel was then subjected to calcination for 3 hours at various temperatures, 400, 500, and 600°C, with 400°C identified as the optimal annealing temperature.

X-ray diffraction analysis confirmed the purity of the ZnO nanoparticles synthesized under optimal conditions, with all peaks corresponding to the hexagonal wurtzite structure of ZnO.

Transmission electron microscopy revealed that the average particle size of the ZnO nanoparticles was approximately 13 ± 5 nm, closely matching the crystallite size. This was further supported by scanning electron microscopy, which showed the nanoparticles to be spherical in shape.

This relatively small particle size represents a significant improvement over previous studies using the same precursor (zinc acetate dihydrate) and the sol-gel method, which produced nanoparticles ranging from 25 to 40 nm. The findings from this study suggest that the use of the Taguchi design can effectively lead to the production of smaller nanoparticles.

Mehr et al. [60] reported a cost-effective and environmentally friendly technique for synthesizing ZnO nanoparticles using *Ferulago angulata* (Schlecht.) Boiss extract as a mild, non-toxic reducing agent and efficient stabilizer, without the need for additional surfactants.

Ferulago angulata (Schlecht.) Boiss, belonging to the Apiaceae family, is a medicinal plant native to western and southwestern Iran, where it is locally known as "Chavil." In traditional medicine, various parts of this plant are used for their diverse therapeutic effects, including sedative, tonic, antidiabetic, aphrodisiac, anti-hemorrhoidal, digestive, anti-ulcer, and anti-parasitic benefits. It is also employed in the treatment of snake bites, spleen disorders, and headaches. Studies have shown that the composition of bioactive compounds in this plant can differ significantly based on the plant part used, the species, or environmental conditions in which it grows [61].

In the aforementioned study [60], besides the *Ferulago angulata* (Schlecht.) Boiss extract, zinc acetate dihydrate was used as the starting material. After performing centrifugation and re-dispersion processes three times, the sample was dried in a hot air oven at 80°C for 6 hours to obtain brown-colored powders and finally calcined at 500°C for 4 hours.

For preparing the plant extract, the plant material was washed and dried in the dark. After, distilled water was added, and the mixture was boiled for 5 minutes. The resulting material, which was yellowish-brown in color, was passed through a wad of cotton and filtered. The aqueous extract was stored at 4°C for further use in the synthesis.

The formation of hexagonal zinc oxide was confirmed by powder X-ray diffraction spectroscopy. Field emission scanning electron microscopy images of the ZnO nanoparticles revealed spheroidal and polydisperse particles, with sizes ranging between 32 and 36 nm.

The photocatalytic activity of the biosynthesized ZnO nanoparticles was evaluated by the degradation of RhB under visible light irradiation, indicating their great potential for photocatalytic applications.

3.4 Titanium dioxide, TiO_2, nanoparticles

Titanium dioxide, an n-type semiconductor with a wide energy band gap, is well-known for its exceptional optical transparency, high refractive index, and chemical stability [62]. It exists in several polymorphic forms, including rutile, anatase, and brookite, with rutile and anatase being the most commonly produced due to their stability. While brookite and anatase are metastable, they can transform into the more stable rutile phase at elevated temperatures [63].

TiO_2 also boasts unique properties such as cost-effectiveness, ease of production, non-toxicity, and strong resistance to chemical corrosion, making it highly suitable for applications in solar cells, chemical sensors, and environmental purification. TiO_2 nanoparticles possess electrical, magnetic, and optical characteristics that differ significantly from their bulk counterparts. Their biocompatibility and non-toxicity make them valuable for biomedical uses, including bone tissue engineering and pharmaceutical applications [64]

Yang et al. [65] reported the synthesis of uniform nanocrystalline TiO_2 via the sol-gel method, using titanium(IV) butoxide and anhydrous alcohol. After aging the mixture at room temperature for 24 hours, it was filtered and washed several times with deionized water and anhydrous alcohol. The drying process took place at 100°C for 12 hours, and calcination was performed at 500°C and 650°C for 2 hours.

X-ray diffraction analysis revealed that the sample sintered at 500°C consisted of pure anatase, while calcination at the higher temperature promoted the formation of both anatase and rutile phases.

Transmission electron microscopy showed that the anatase TiO_2 nanoparticles had a spherical morphology, with an average diameter of approximately 20 nm. To evaluate the photocatalytic activity of the as-prepared TiO_2 nanoparticles, the photocatalytic oxidation of methyl orange in TiO_2 suspension under UV illumination was investigated.

The experiments demonstrated that the methyl orange solution was significantly degraded in the aqueous TiO_2 suspension, and the photocatalytic activity of TiO_2 containing the rutile phase was found to be inferior to that of pure anatase TiO_2 powder.

In a study by Subhapriya et al. [64], titanium dioxide nanoparticles were biosynthesized using leaf extract from *Trigonella foenum-graecum* (commonly known as fenugreek).

Fenugreek, a leguminous aromatic plant, is widespread in regions including the Middle East, Asia, and Europe [64]. Traditionally, it has been used for its medicinal properties, such as lowering blood sugar and cholesterol levels in both diabetic and non-diabetic individuals, as well as for its antioxidant and antibacterial activities. The plant contains bioactive compounds like alkaloids, flavonoids, steroids, and saponins, which contribute to its long-standing use in traditional medicine. Fenugreek has been identified as a promising medicinal plant with potential applications in treating diseases and serving as a raw material in the pharmaceutical industry, particularly in the production of steroidal hormones [66].

In the aforementioned research [64], TiO_2 nanoparticles were synthesized using titanium oxy sulfate, sodium hydroxide, and fenugreek leaf extract. The resulting precipitate was washed, filtered, and sintered at 700°C for three hours. To prepare the plant extract, fenugreek leaves were first washed with distilled water, then weighed and ground using a mortar and pestle. The crushed leaves were mixed with distilled water, stirred, and filtered to obtain the extract for use in the synthesis process.

X-ray diffraction analysis confirmed the formation of pure anatase TiO_2, while high-resolution scanning electron microscopy revealed that the nanoparticles were spherical, with sizes ranging from 20 to 90 nm, including some aggregates. The synthesized TiO_2 nanoparticles demonstrated strong antimicrobial activity against several microorganisms, including *Yersinia enterocolitica*, *Escherichia coli*, *Staphylococcus aureus*, and *Enterococcus faecalis*.

According to the authors, this method offers an eco-friendly and cost-effective approach, with potential applications in drug delivery, the treatment of infectious diseases and cancer, as well as commercial, sensor, electronic, and medical fields.

3.5 Magnesium oxide, MgO, nanoparticles

Magnesium oxide is a wide band gap semiconductor with a band gap *of approximately* 7 eV [67, 68].

Although MgO crystallizes in the rocksalt structure [69] and is often considered an almost ideal insulating ionic solid [70], it exhibits some characteristics typical of semiconductors, such as a wide valence band, a large dielectric constant, and a low exciton binding energy—values that differ from those of typical insulators. For these and other reasons, MgO can be regarded as a semiconductor [69].

Among inorganic nanoparticles, MgO has garnered considerable attention due to its fundamental properties, such as a high specific surface area and reactivity, making it valuable in applications like catalysis, ceramics, and environmental remediation. Additionally, its thermodynamic stability, low dielectric constant, refractive index, and wide bandgap make MgO relevant in electrochemical applications. Furthermore, its relatively non-toxic properties have been recognized as safe for human use by the United States Food and Drug Administration [68].

Researchers have focused on the synthesis of MgO nanoparticles because of their distinctive properties. These include a high ionic nature, a simple crystal structure and stoichiometry, a large specific surface area, and high reactivity due to numerous active sites. Additionally, the size and shape of the particles can be tailored as required, and the surface contains numerous crystal defects [69].

Salman et al. [71] synthesized MgO nanoparticles using the sol-gel method. Magnesium nitrate hexahydrate and sodium hydroxide were employed, with distilled water serving as the solvent and ethanol used as a washing reagent. The precipitate obtained was filtered and washed several times to yield the final product. The material was then placed in a vacuum oven at 80°C for 4 hours to remove moisture and subsequently calcined at 400°C for 3 hours. X-ray diffraction analysis confirmed the presence of a pure MgO phase, while field emission scanning electron microscopy images revealed spherical particles with an approximate size of 50 nm.

The specific surface area of the MgO nanoparticles was measured using the Brunauer-Emmett-Teller method and found to be approximately 191 m^2/g.

Umaralikhan et al. [72] reported the synthesis of MgO nanoparticles using aqueous leaf extracts of *Psidium guajava* and *Aloe vera*.

Psidium guajava belongs to the Myrtaceae family and is native to Mexico, though it is now distributed across South America, Europe, Africa, and Asia [72, 73]. It is a small tree that can grow up to 10 meters tall, featuring thin, smooth bark that peels in patches. The leaves are the opposite, having short stalks and are oval in shape with noticeable pinnate veins, ranging from 5 to 15 cm in length.

Psidium guajava is commonly referred to as "guava" in English; "goyave" or "goyavier" in French; "Guavenbaum" or "Guayave" in German; "banjiro" in Japanese; "goiaba" or "goiabeiro" in Portuguese; "araçá-goiaba," "araçá-guaçú," or "guaiaba" in Portuguese from Brazil; and "guayaba" or "guayabo" in Spanish [73].

It is an important food crop and medicinal plant widely used in food and traditional medicine across the world [72]. *Psidium guajava* is widely recognized for its antispasmodic and antimicrobial effects, particularly in the treatment of diarrhea and dysentery. It has also been commonly used as a hypoglycemic agent. Numerous pharmacological studies have confirmed the plant's various properties, including antioxidant, hepatoprotective, anti-allergic, antimicrobial, antigenotoxic, antiplasmodial, cytotoxic, cardioactive, anticough, antidiabetic, anti-inflammatory, and antinociceptive activities, all of which validate its traditional medicinal applications [72, 73].

Aloe vera, a member of the Xanthorrhoeaceae family, is a perennial, succulent, xerophytic plant with crassulacean acid metabolism. Originally from the western Arabian Peninsula, it is now widely distributed and cultivated worldwide. *Aloe vera* has been appreciated for its medicinal and healing properties for centuries. The inner gel contains more than 75 active compounds, although the precise therapeutic effects of each have yet to be fully understood. Traditionally, *Aloe vera* has been used to treat a variety of conditions and is commonly included in dietary supplements and food products. Its gel is frequently incorporated into skincare products like lotions and sunscreens. Many of the health benefits associated with *Aloe vera* extracts are believed to stem from the polysaccharides present in the inner leaf's parenchymatous tissue [72, 74, 75].

In the aforementioned study [72], magnesium nitrate dissolved in distilled water and the plant extracts were used as the starting materials. After the supernatant solvent was evaporated, the resulting product was ground into a fine powder and purified at 600°C for 2 hours.

To prepare the leaf extracts, fresh leaves of *Psidium guajava* and *Aloe vera* were first cleaned under running tap water and then in distilled water. The leaves were then added to distilled water and boiled at 40–50°C for 5 minutes. The extract was filtered, and the filtrate was collected and stored at room temperature for later use.

X-ray diffraction analysis confirmed the high purity of the samples prepared via this method, with all diffraction peaks corresponding to cubic MgO, and field emission scanning electron microscopy images revealed that the MgO nanoparticles ranged in size from 50 to 90 nm.

The antibacterial activity of the MgO nanoparticles was tested against Gram-negative *E. coli* and Gram-positive *S. aureus* using the disk diffusion method, demonstrating significant and similar effects against both species.

The authors concluded that the green-synthesized nanoparticles hold promising potential for antimicrobial applications, with both plant extracts demonstrating excellent suitability for nanoparticle synthesis.

3.6 Nickel oxide, NiO, nanoparticles

Nickel oxide is a p-type semiconductor and is considered a model semiconductor due to its wide bandgap energy of 3.6–4 eV [76].

NiO nanoparticles have been widely investigated because of their mechanical, electronic, magnetic, and optical properties, and they have applications in various fields, such as catalysis, electrochromic films, and fuel cell electrodes [77].

According to Manna et al. [78], bulk NiO exhibits antiferromagnetic behavior below the Néel temperature of 523 K. However, the magnetic properties of NiO can be tailored by reducing its size to the nanometer scale. Specific magnetic phases, such as superparamagnetic, superantiferromagnetic, and ferromagnetic ordering, appear depending on the size, shape, and synthesis route of the nanostructured NiO.

Alagiri et al. [77] synthesized NiO nanoparticles using the sol-gel method. The materials used included nickel(II) nitrate hexahydrate and agarose polysaccharide. The obtained gel was heated at 100°C for 4 hours, and the dried material was calcined at 400°C for 3 hours.

The X-ray diffraction pattern of the sample matched well with the Bragg reflections of standard face-centered cubic NiO, and the transmission electron microscopy images revealed that the synthesized product consisted of spherical nanoparticles approximately 3 nm in size.

The magnetic characterization showed that the as-prepared nanoparticles exhibited superparamagnetic behavior at room temperature.

Gebretinsae et al. [76] reported, for the first time, the synthesis of NiO nanoparticles via a green chemistry process using *Opuntia ficus-indica* extract.

Opuntia ficus-indica, commonly known as cactus, belongs to the Cactaceae family. Although native to Mexico, it grows wild in arid and semi-arid regions of South and Central America, Africa, and the Mediterranean [79]. This prickly pear cactus has garnered interest worldwide because it thrives where few other crops can, such as in Ethiopia, where successful crop growth is tremendously limited [80]. The cactus is primarily cultivated for its fruit [81], but it also has healthcare uses due to its high content of polyphenols and its antioxidant, anti-inflammatory, and anxiolytic properties [80].

The cladodes of *Opuntia ficus-indica*, due to their chemical composition, exhibit antioxidant activity, with bioactive compounds like phenolics, flavonoids, alkaloids, aspalathin, thiamine, riboflavin, and niacin acting as potent reducing and capping agents [76].

In the aforementioned work [76], the starting materials were nickel(II) nitrate hexahydrate and cactus cladode extract, prepared from freshly collected cladodes, coincidentally, in Ethiopia. The final solution was dried at 80°C, and the obtained material was annealed in air for 2 hours at 500°C.

To prepare the extract, the dried cladodes were crushed into a green paste-like solution, which was then centrifuged. After adding distilled water, the mixture was heated at 60°C for 45 minutes. Finally, the extract was filtered three times and stored at room temperature for 24 hours.

X-ray diffraction analysis showed the formation of highly pure NiO powders, and scanning electron microscopy revealed that the nanoparticles were spherical, with sizes ranging from 20 to 35 nm. However, some degree of agglomeration was observed, which the authors attributed to the high surface energy and surface tension of the NiO nanoparticles.

4. Future trends and challenges

In recent years, there has been a growing focus on the synthesis of nanoscale metals through chemical, physical, and environmentally friendly methods [43, 82].

However, green synthesis approaches are gradually replacing traditional physical and chemical methods [83]. This shift is driven by concerns over the high energy demands [84, 85], the release of harmful chemicals [83–85], and the complexity of equipment and synthesis conditions associated with conventional methods [43].

The green synthesis utilizes natural, eco-friendly materials, mainly as biological reducing agents [43]. Some of these green substances can function both as capping agents [86] and dispersants, which not only minimizes energy consumption but also eliminates the need for harmful and toxic chemicals [43, 87]. Currently, green synthesis techniques mainly use microorganisms (such as fungi, bacteria, and algae) or extracts from plant parts, including leaves, flowers, roots, peels, barks, and fruits [43, 88, 89].

The production of nanomaterials, including nanoparticles and nanostructures, using plants and plant-derived extracts is referred to as phytogenic synthesis [88]. This method is based on the concept of phytosynthesis, which utilizes various secondary metabolites found in different plant parts for the eco-friendly creation of nanoparticles [88, 90].

Green materials, such as those containing polyphenols and proteins, can act as natural reducing agents, replacing traditional chemical reagents to reduce metal ions to lower valence states. When exposed to optimal conditions, such as the right temperature, concentration, and air exposure, metal oxide nanoparticles can be produced through this method [43, 82].

This green synthesis approach has gained popularity among researchers due to its sustainability and the abundant availability of plant extracts, which makes it a highly promising avenue for nanomaterial fabrication [88, 91]. In general, the process involves collecting fresh plant material, drying it, and grinding it into a powder. The powder is then boiled in water to activate the phytochemicals present in the plant, which contribute to the biosynthesis of nanoparticles [88].

Significant research into the synthesis of nanomaterials from natural sources has led to the development of numerous green and biogenic processes. These "green technologies" are gradually transitioning from laboratory experiments to commercial and industrial applications, though they face several challenges along the way. Current green synthesis methods emphasize the use of natural resources, safe solvents, efficient processes, and even biological systems' metabolic functions [88, 92]. Additionally, it has been noted that green nanomaterials often retain inherited structures, which are functional groups from their organic precursors. These structures give nanomaterials unique mechanical, chemical, electrical, and optical properties, making them useful in diverse areas, from biomedical applications to electronics [88]. Under certain conditions, the quality of green-synthesized metal nanoparticles even surpasses those synthesized by chemical methods [43].

Recent research trends suggest that green nanomaterials, particularly those derived from biomass or natural precursors, will likely drive the commercialization of nanoscale composites, self-assemblies, and new structural forms. As a result, there will be an increase in studies focused on producing green nanoparticles with properties tailored for specific applications [88]. Moreover, the involvement of biomass waste in the green synthesis process can facilitate the transition of nanoparticle production to the principles of a circular economy [93].

However, the environmental and health aspects of using green or biomass-derived nanomaterials and nanostructures need to be investigated in detail before considering their industrial-scale production and commercial-scale consumption [88]. Despite the numerous advantages attributed to the green synthesis of nanoparticles and nanomaterials, particularly those derived from plant extracts, it is crucial to examine the synthesis methods and source materials used for these green nanomaterials before exploring their physicochemical properties and applications. Also, whenever a novel material is produced, its potential for harmful and toxicological interactions with humans and animals must be considered [88].

5. Conclusions

Nanoparticles play a crucial role in nanotechnology, with their applications growing quickly because of their distinctive optical, chemical, electrical, and mechanical characteristics.

In particular, metal oxide nanoparticles are widely studied, as demonstrated by an analysis of papers published in SCOPUS over the last ten years. This confirms the growing interest in the synthesis of metal oxide nanoparticles, largely due to the abundance of their natural precursors, environmental friendliness, affordability, and favorable electrical and optical properties. In some cases, they also exhibit chemical stability.

In recent decades, significant efforts have been devoted to developing reliable synthetic methods for nanosized materials, enabling precise control over their size, shape, and other unique properties. Among the bottom-up approaches, the sol-gel technique has been widely utilized, offering benefits such as simplicity, low cost, lower processing temperatures, and precise control over nanoparticle size and shape.

However, in recent years, the biosynthesis of nanoparticles using eco-friendly techniques—replacing hazardous and/or expensive chemicals with plants, algae, bacteria, fungi, and viruses—has gained increasing attention.

The growing number of reports on the synthesis of metal oxide nanoparticles using green approaches, specifically publications involving plant extracts, supports the advantages attributed to them, namely their availability, safety, and capacity to produce large quantities of nanoparticles. Additionally, these extracts can serve as both reducing and stabilizing agents in the synthesis process.

This mini-review focuses on the synthesis of six different metal oxide nanoparticles, CuO, Cu_2O, ZnO, TiO_2, MgO, and NiO, using both the conventional sol-gel method and a green approach based on plant extracts.

It has been demonstrated that various metal oxide nanoparticles can be synthesized through a green sol-gel route, with starting materials such as isopropyl alcohol, ethanolamine, ethylene glycol, oxalic acid, or agarose polysaccharide being substituted by plant extracts such as *Yucca gigantea*, *Phoenix dactylifera L.*, *Ferulago angulata (Schlecht.) Boiss*, *Trigonella foenum-graecum*, *Psidium guajava*, *Aloe vera*, and *Opuntia ficus-indica*.

These extracts are easy and safe to obtain, without the need for hazardous chemicals or expensive, elaborate equipment. Using these green routes, the synthesis of pure metal oxide nanoparticles with average particle sizes ranging from 20 to 90 nm has been reported.

Acknowledgements

This research is sponsored by national funds through FCT—Fundação para a Ciência e Tecnologia—under projects UIDB/00285/2020 and LA/P/0112/2020.

Conflict of interest

The author declares no conflict of interest.

Author details

Susana Devesa
CEMMPRE, ARISE, Department of Mechanical Engineering, University of Coimbra, Coimbra, Portugal

*Address all correspondence to: sdevesa@dem.uc.pt

References

[1] Boholm M. The use and meaning of nano in American English: Towards a systematic description. Ampersand. 2016;**3**:163-173

[2] Buzea C, Pacheco II, Robbie K. Nanomaterials and nanoparticles: Sources and toxicity. Biointerphases. 2007;**2**(4):17-71

[3] Joudeh N, Linke D. Nanoparticle classification, physicochemical properties, characterization, and applications: A comprehensive review for biologists. Journal of Nanobiotechnology. 2022;**20**(1):262

[4] Srivastava S, Bhargava A. Green nanotechnology: An overview. In: Green Nanoparticles: The Future of Nanobiotechnology. Singapore: Springer; 2022. pp. 1-13

[5] Fanfair D, Desai S, Kelty C. The early history of nanotechnology. Connexions Connexions. 2007;**5**:1-15

[6] Ullah Khan S, Saleh TA, Wahab A, Ullah Khan MH, Khan D, Ullah Khan W, et al. Nanosilver: New ageless and versatile biomedical therapeutic scaffold. International Journal of Nanomedicine. 2018;**13**:733-762

[7] Papazoglou ES, Parthasarathy A. Introduction. In: Bionanotechnology. Vol. 7. San Rafael, California (USA): Morgan & Claypool Publisherspp; 2007. 1-14

[8] Aigbe UO, Osibote OA. Green synthesis of metal oxide nanoparticles, and their various applications. Journal of Hazardous Materials Advances. 2024;**13**: 100401

[9] Szczyglewska P, Feliczak-Guzik A, Nowak I. Nanotechnology–general aspects: A chemical reduction approach to the synthesis of nanoparticles. Molecules. 2023;**28**(13):4932

[10] Auffan M, Rose J, Bottero JY, Lowry GV, Jolivet JP, Wiesner MR. Towards a definition of inorganic nanoparticles from an environmental, health and safety perspective. Nature Nanotechnology. 2009;**4**(10):634-641

[11] Borm PJ, Robbins D, Haubold S, Kuhlbusch T, Fissan H, Donaldson K, et al. The potential risks of nanomaterials: A review carried out for ECETOC. Particle and Fibre Toxicology. 2006;**3**(1):11

[12] International Standard. ISO 80004-1: 2023 Nanotechnologies—Vocabulary Part 1: Core Vocabulary. Geneva, Switzerland: International Organization for Standardization; 2023

[13] Khan I, Saeed K, Khan I. Nanoparticles: Properties, applications and toxicities. Arabian Journal of Chemistry. 2019;**12**(7):908-931

[14] Hamida RS, Ali MA, Abdelmeguid NE, Al-Zaban MI, Baz L, Bin-Meferij MM. Lichens—A potential source for nanoparticles fabrication: A review on nanoparticles biosynthesis and their prospective applications. Journal of Fungi. 2021;7(4):291

[15] Anu Mary Ealia S, Saravanakumar MP. A review on the classification, characterisation, synthesis of nanoparticles and their application. IOP Conference Series: Materials Science and Engineering. 2017;**263**:032019

[16] Ijaz I, Gilani E, Nazir A, Bukhari A. Detail review on chemical, physical and green synthesis, classification, characterizations and applications of nanoparticles. Green Chemistry Letters and Reviews. 2020;**13**(3):223-245

[17] Sajid M, Płotka-Wasylka J. Nanoparticles: Synthesis, characteristics, and applications in analytical and other sciences. Microchemical Journal. 2020; **154**:104623

[18] Khalil M, Jan BM, Tong CW, Berawi MA. Advanced nanomaterials in oil and gas industry: Design, application and challenges. Applied Energy. 2017; **191**:287-310

[19] Nunes D, Pimentel A, Gonçalves A, Pereira S, Branquinho R, Barquinha P, et al. Metal oxide nanostructures for sensor applications. Semiconductor Science and Technology. 2019;**34**(4): 043001

[20] Bin WH, Chen JS, Hng HH, Wen (David) Lou X. Nanostructured metal oxide-based materials as advanced anodes for lithium-ion batteries. Nanoscale. 2012;**4**(8):2526

[21] Devan RS, Patil RA, Lin J, Ma Y. One-dimensional metal-oxide nanostructures: Recent developments in synthesis, characterization, and applications. Advanced Functional Materials. 2012;**22**(16):3326-3370

[22] Shahid M, Sagadevan S, Ahmed W, Zhan Y, Opaprakasit P. Metal oxides for optoelectronic and photonic applications: A general introduction. In: Metal Oxides for Optoelectronics and Optics-Based Medical Applications. Amsterdam, The Netherlands: Elsevier; 2022. pp. 3-31

[23] Walia S, Balendhran S, Nili H, Zhuiykov S, Rosengarten G, Wang QH, et al. Transition metal oxides – Thermoelectric properties. Progress in Materials Science. 2013;**58**(8): 1443-1489

[24] Moumen A, Kumarage GCW, Comini E. P-type metal oxide semiconductor thin films: Synthesis and chemical sensor applications. Sensors. 2022;**22**(4):1359

[25] Wang C, Yin L, Zhang L, Xiang D, Gao R. Metal oxide gas sensors: Sensitivity and influencing factors. Sensors. 2010;**10**(3):2088-2106

[26] Guo T, Yao MS, Lin YH, Nan CW. A comprehensive review on synthesis methods for transition-metal oxide nanostructures. **CrystEngComm**. 2015; **17**(19):3551-3585

[27] Dębek P, Feliczak-Guzik A, Nowak I. Nanostructures: General information. The use of nanoobjects in medicine and cosmetology. Advances in Hygiene and Experimental Medicine. 2017;**71**:1055-1062

[28] Alhalili Z. Metal oxides nanoparticles: General structural description, chemical, physical, and biological synthesis methods, role in pesticides and heavy metal removal through wastewater treatment. Molecules. 2023;**28**(7):3086

[29] Bokov D, Turki Jalil A, Chupradit S, Suksatan W, Javed Ansari M, Shewael IH, et al. Nanomaterial by sol-gel method: Synthesis and application. Advances in Materials Science and Engineering. 2021;**1**:5102014

[30] Prabakaran S, Rajan M. Biosynthesis of nanoparticles and their roles in numerous areas. Comprehensive Analytical Chemistry. 2021;**94**:1-47

[31] Raduwan NF, Shaari N, Kamarudin SK, Masdar MS, Yunus RM. An overview of nanomaterials in fuel cells: Synthesis method and application. International Journal of Hydrogen Energy. 2022;**47**(42):18468-18495

[32] Pierre AC. Introduction to sol-Gel Processing. Vol. 1. Boston, MA: Springer US; 1998

[33] Dimitriev Y, Ivanova Y, Iordanova R. History of sol-gel science

and technology. Journal of the University of Chemical technology and Metallurgy. 2008;**43**(2):181-192

[34] Schmidt H. Considerations about the sol-gel process: From the classical sol-gel route to advanced chemical nanotechnologies. Journal of Sol-Gel Science and Technology. 2006;**40**(2–3): 115-130

[35] Dislich H. Sol-gel: Science, processes and products. Journal of Non-Crystalline Solids. 1986;**80**(1–3):115-121

[36] Sakka S. Sol-gel technology as reflected in journal of sol-gel science and technology. Journal of Sol-Gel Science and Technology. 2003;**26**(1/3):29-33

[37] Parashar M, Shukla VK, Singh R. Metal oxides nanoparticles via sol–gel method: A review on synthesis, characterization and applications. Journal of Materials Science: Materials in Electronics. 2020;**31**(5):3729-3749

[38] Navas D, Fuentes S, Castro-Alvarez A, Chavez-Angel E. Review on sol-gel synthesis of perovskite and oxide nanomaterials. Gels. 2021;**7**(4):275

[39] Soltys L, Olkhovyy O, Tatarchuk T, Naushad M. Green synthesis of metal and metal oxide nanoparticles: Principles of green chemistry and raw materials. Magnetochemistry. 2021;**7**(11):145

[40] Erythropel HC, Zimmerman JB, de Winter TM, Petitjean L, Melnikov F, Lam CH, et al. The green ChemisTREE: 20 years after taking root with the 12 principles. Green Chemistry. 2018; **20**(9):1929-1961

[41] Deena Raj KM, Sujatha S, Chanthini KMP. Green synthesis of CuO nanoparticle using Cyperus rotundus and evaluation of their anticancer potentials against lung cancer cells invitro. Biocatalysis and Agricultural Biotechnology. 2024;**57**:103119

[42] Kerour A, Boudjadar S, Bourzami R, Allouche B. Eco-friendly synthesis of cuprous oxide (Cu_2O) nanoparticles and improvement of their solar photocatalytic activities. Journal of Solid State Chemistry. 2018;**263**:79-83

[43] Miu BA, Dinischiotu A. New green approaches in nanoparticles synthesis: An overview. Molecules. 2022;**27**(19): 6472

[44] Sunny NE, Mathew SS, Chandel N, Saravanan P, Rajeshkannan R, Rajasimman M, et al. Green synthesis of titanium dioxide nanoparticles using plant biomass and their applications- a review. Chemosphere. 2022;**300**:134612

[45] Alotaibi B, Elekhnawy E, El-Masry TA, Saleh A, El-Bouseary MM, Alosaimi ME, et al. Green synthetized Cu-oxide nanoparticles: Properties and applications for enhancing healing of wounds infected with Staphylococcus aureus. International Journal of Pharmaceutics. 2023;**645**:123415

[46] Kumar N, Parui SS, Limbu S, Mahato DK, Tiwari N, Chauhan RN. Structural and optical properties of sol–gel derived CuO and Cu_2O nanoparticles. Materials Today Proceedings. 2021;**41**: 237-241

[47] Yang C, Xiao F, Wang J, Su X. Synthesis and microwave modification of CuO nanoparticles: Crystallinity and morphological variations, catalysis, and gas sensing. Journal of Colloid and Interface Science. 2014;**435**:34-42

[48] Chavali MS, Nikolova MP. Metal oxide nanoparticles and their applications in nanotechnology. SN Applied Sciences. 2019;**1**(6):607

[49] Ighalo JO, Sagboye PA, Umenweke G, Ajala OJ, Omoarukhe FO, Adeyanju CA, et al. CuO nanoparticles

(CuO NPs) for water treatment: A review of recent advances. Environmental Nanotechnology, Monitoring & Management. 2021;**15**:100443

[50] Wang F, Li H, Yuan Z, Sun Y, Chang F, Deng H, et al. A highly sensitive gas sensor based on CuO nanoparticles synthetized via a sol–gel method. RSC Advances. 2016;**6**(83): 79343-79349

[51] Patel S. Yucca: A medicinally significant genus with manifold therapeutic attributes. Natural Products and Bioprospecting. 2012;**2**(6):231-234

[52] Kumari N, Kumari P, Jha AK, Prasad K. Green synthesis of Cu_2O nanoparticles using grape juice and its antimicrobial activity. AIP Conference Proceedings. 2020;**2220**(1):020042-1-020042-4

[53] Zayyoun N, Bahmad L, Laânab L, Jaber B. The effect of pH on the synthesis of stable Cu_2O/CuO nanoparticles by sol–gel method in a glycolic medium. Applied Physics A. 2016;**122**(5):488

[54] Berra D, Laouini SE, Benhaoua B, Ouahrani MR, M. R. BD. Green synthesis of copper oxide nanoparticles by *Pheonix dactylifera* L leaves extract. Digest Journal of Nanomaterials and Biostructures. 2018;**13**(4):1231-1238

[55] Chao CT, Krueger RR. The date palm (*Phoenix dactylifera* L.): Overview of biology, uses, and cultivation. Hort Science. 2007;**42**(5):1077-1082

[56] Baliga MS, Baliga BRV, Kandathil SM, Bhat HP, Vayalil PK. A review of the chemistry and pharmacology of the date fruits (*Phoenix dactylifera* L.). Food Research International. 2011;**44**(7):1812-1822

[57] Ong CB, Ng LY, Mohammad AW. A review of ZnO nanoparticles as solar photocatalysts: Synthesis, mechanisms and applications. Renewable and Sustainable Energy Reviews. 2018;**81**: 536-551

[58] Uikey P, Vishwakarma K. Review of zinc oxide (ZnO) nanoparticles applications and properties. International Journal of Emerging Technology in Computer Science & Electronics. 2016;**21**(2):239-242

[59] Chung YT, Ba-Abbad MM, Mohammad AW, Hairom NHH, Benamor A. Synthesis of minimal-size ZnO nanoparticles through sol–gel method: Taguchi design optimisation. Materials and Design. 2015;**87**:780-787

[60] Shayegan Mehr E, Sorbiun M, Ramazani A, Taghavi FS. Plant-mediated synthesis of zinc oxide and copper oxide nanoparticles by using *Ferulago angulata* (*schlecht*) boiss extract and comparison of their photocatalytic degradation of Rhodamine B (RhB) under visible light irradiation. Journal of Materials Science: Materials in Electronics. 2018;**29**(2): 1333-1340

[61] Hazrati S, Ebadi MT, Mollaei S, Khurizadeh S. Evaluation of volatile and phenolic compounds, and antioxidant activity of different parts of *Ferulago angulata* (*schlecht.*) Boiss. Industrial Crops and Products. 2019;**140**:111589

[62] Zhao Y, Li C, Liu X, Gu F, Jiang H, Shao W, et al. Synthesis and optical properties of TiO_2 nanoparticles. Materials Letters. 2007;**61**(1):79-83

[63] Rajaram P, Jeice AR, Jayakumar K. Review of green synthesized TiO_2 nanoparticles for diverse applications. Surfaces and Interfaces. 2023;**39**:102912

[64] Subhapriya S, Gomathipriya P. Green synthesis of titanium dioxide (TiO_2) nanoparticles by

Trigonella foenum-graecum extract and its antimicrobial properties. Microbial Pathogenesis. 2018;**116**:215-220

[65] Yang H, Zhang K, Shi R, Li X, Dong X, Yu Y. Sol–gel synthesis of TiO_2 nanoparticles and photocatalytic degradation of methyl orange in aqueous TiO_2 suspensions. Journal of Alloys and Compounds. 2006;**413**(1–2):302-306

[66] Kor ZM, Zadeh JB. Fenugreek (*Trigonella foenum-graecum* L.) as a valuable medicinal plant. International Journal of Advanced Biological and Biomedical Research. 2013;**1**(8):922-931

[67] Das A, Mandal AC, Roy S, Nambissan PMG. Internal defect structure of calcium doped magnesium oxide nanoparticles studied by positron annihilation spectroscopy. AIP Advances. 2018;**8**(9):095013-1-095013-13

[68] Tharani K, Jegatha Christy A, Sagadevan S, Nehru LC. Fabrication of magnesium oxide nanoparticles using combustion method for a biological and environmental cause. Chemical Physics Letters. 2021;**763**:138216

[69] Wayan Sutapa I, Wahid Wahab A, Taba P, La Nafie N. Synthesis and structural profile analysis of the MgO nanoparticles produced through the sol-gel method followed by annealing process. Oriental Journal of Chemistry. 2018;**34**(2):1016-1025

[70] Canney SA, Sashin VA, Ford MJ, Kheifets AS. Electronic band structure of magnesium and magnesium oxide: Experiment and theory. Journal of Physics: Condensed Matter. 1999;**11**(39): 7507-7522

[71] Salman KD, Abbas HH, Aljawad HA. Synthesis and characterization of MgO nanoparticle via microwave and sol-gel methods. Journal of Physics Conference Series. 2021;**1973**(1):012104

[72] Umaralikhan L, Jaffar JM, M. Green synthesis of MgO nanoparticles and it antibacterial activity. Iranian Journal of Science and Technology, Transactions A: Science. 2018;**42**(2):477-485

[73] Gutiérrez RMP, Mitchell S, Solis RV. Psidium guajava: A review of its traditional uses, phytochemistry and pharmacology. Journal of Ethnopharmacology. 2008;**117**(1):1-27

[74] Shekhawat MS. Microstructural and histochemical variations during in vitro to in vivo plant developments in *Aloe vera* (L.) Burm.f (Xanthorrhoeaceae). Industrial Crops and Products. 2021;**160**: 113162

[75] Minwuyelet T, Sewalem M, Gashe M. Review on therapeutic uses of Aloe vera. Global Journal of Pharmacology. 2017;**11**(2):14-20

[76] Gebretinsae HG, Tsegay MG, Nuru ZY. Biosynthesis of nickel oxide (NiO) nanoparticles from cactus plant extract. Materials Today Proceedings. 2021;**36**:566-570

[77] Alagiri M, Ponnusamy S, Muthamizhchelvan C. Synthesis and characterization of NiO nanoparticles by sol–gel method. Journal of Materials Science: Materials in Electronics. 2012; **23**(3):728-732

[78] Manna S, Deb AK, Jagannath J, De SK. Synthesis and room temperature ferromagnetism in Fe doped NiO Nanorods. The Journal of Physical Chemistry C. 2008;**112**(29):10659-10662

[79] Aragona M, Lauriano ER, Pergolizzi S, Faggio C. *Opuntia ficus - indica* (L.) miller as a source of

bioactivity compounds for health and nutrition. Natural Product Research. 2018;**32**(17):2037-2049

[80] Silva MA, Albuquerque TG, Pereira P, Ramalho R, Vicente F, Oliveira MBPP, et al. *Opuntia ficus-indica* (L.) Mill.: A multi-benefit potential to be exploited. Molecules. 2021;**26**(4):951

[81] de Cortázar VG, Nobel PS. Biomass and fruit production for the prickly pear cactus, Opuntia ficus-indica. Journal of the American Society for Horticultural Science. 1992;**117**(4):558-562

[82] Saiqa, Ikram SA. Silver nanoparticles: One pot green synthesis using Terminalia Arjuna extract for biological application. Journal of Nanomedicine and Nanotechnology. 2015:**6**(4):1-6

[83] Varma RS. Journey on greener pathways: From the use of alternate energy inputs and benign reaction media to sustainable applications of nano-catalysts in synthesis and environmental remediation. Green Chemistry. 2027; **2014**:*16*

[84] Chahardoli A, Karimi N, Fattahi A. Nigella arvensis leaf extract mediated green synthesis of silver nanoparticles: Their characteristic properties and biological efficacy. Advanced Powder Technology. 2018;**29**:202-210

[85] Kumar V, Yadav SK. Plant-mediated synthesis of silver and gold nanoparticles and their applications. Journal of Chemical Technology & Biotechnology. 2009;**84**:151-157

[86] Tiwari S, Verma SK, Bhagat P, Yadav S, Sharma R, Aseri GK, et al. An overview of the phytosynthesis of various metal nanoparticles. 3 Biotech. 2021;**11**:478

[87] Mustapha T, Misni N, Ithnin NR, Daskum AM, Unyah NZ. A review on plants and microorganisms mediated synthesis of silver nanoparticles, role of plants metabolites and applications. International Journal of Environmental Research and Public Health. 2022;**19**:674

[88] Arif MD, Hoque ME, Rahman MZ, Shafoyat MU. Emerging directions in green nanomaterials: Synthesis, physicochemical properties and applications. Materials Today Communications. 2024;**40**:109335

[89] Dhand V, Soumya L, Bharadwaj S, Chakra S, Bhatt D, Sreedhar B. Green synthesis of silver nanoparticles using Coffea Arabica seed extract and its antibacterial activity. Materials Science and Engineering: C. 2016;**58**:36-43

[90] Devi HS, Boda MA, Shah MA, Parveen S, Wani AH. Green synthesis of iron oxide nanoparticles using platanus orientalis leaf extract for antifungal activity. Green Processing and Synthesis. 2019;**8**:38-45

[91] Hoag GE, Collins JB, Holcomb JL, Hoag JR, Nadagouda MN, Varma RS. Degradation of bromothymol blue by 'greener' nano-scale zero-valent iron synthesized using tea polyphenols. Journal of Materials Chemistry. 2009;**19**:8671

[92] Alsammarraie FK, Wang W, Zhou P, Mustapha A, Lin M. Green synthesis of silver nanoparticles using turmeric extracts and investigation of their antibacterial activities. Colloids and Surfaces. B, Biointerfaces. 2018;**171**: 398-405

[93] Ying S, Guan Z, Ofoegbu PC, Clubb P, Rico C, He F, et al. Green synthesis of nanoparticles: Current developments and limitations. Environmental Technology and Innovation. 2022;**26**:102336

Chapter 2

Application of Silica Sol-Gel Coatings for the Bioactivation of Metallic Implants

Francisco Romero-Gavilán and Julio Suay

Abstract

Metallic implants can cause problems due to the release of corrosion products that complicate the successful integration of the material in the human body. Additionally, some of these materials are considered bioinert. The versatility of the sol-gel technique allows tuning the properties of silicon networks that can be applied as coatings on metallic implants. These coatings can protect metal implants from corrosion problems, and they can increase their bioactivity. The possibility of adjusting the sol-gel coating biodegradation rate also makes them a very good option to be used as release vehicles. In addition to the bioactivation generated by the coating itself, bioactive compounds can be released from the sol-gel material promoting the healing process or preventing infections. This chapter will describe the main advances in the development and characterization of these silica sol-gel coatings published in the literature.

Keywords: dental implant, osseointegration, hybrid, Si, alkoxysilane

1. Introduction

Metal implants are widely used in biomedical applications due to their excellent mechanical properties and biocompatibility. Common materials include stainless steel (SS), pure titanium and its alloys, and cobalt-based alloys [1]. Stainless steel is frequently employed in orthopedic implants (*e.g.* intramedullary nails, plates, and screws for bone fixation, and joint prostheses), dental implantology, and cardiovascular devices like coronary stents and heart valves [2]. AISI 316 L stainless steel is particularly notable due to its high molybdenum (2–3%) and chromium (17–20%) content and low carbon content (less than 0.03%), which enhance its corrosion resistance. Additionally, this steel is known for its good ductility, work hardening capacity, and fatigue properties [3]. Compared to other metallic biomaterials, stainless steel is cost-effective, making it popular in developing countries [4]. However, its higher modulus of elasticity compared to bone can cause mechanical incompatibility issues. Its lack of biofunctionality further limits its use, and it is prone to corrosion in low-oxygen regions or areas with cracks, which can lead to allergic reactions and inflammation due to the toxicity of nickel and chromium [1].

Titanium and its alloys are widely used in implants due to their excellent biocompatibility, corrosion resistance, and favorable mechanical properties. These materials are commonly used in dental implants, fracture fixation devices (plates and screws), and craniofacial and maxillofacial implants [5]. Their lower modulus of elasticity compared to stainless steel makes them more compatible with bone tissue. However, there are concerns about the potential release of metal ions and the toxicity associated with vanadium in the Ti6Al4V alloy [6]. The excellent biocompatibility of titanium alloys is attributed to the formation of stable oxides on their surfaces, rendering them bioinert [7]. However, once the TiO_2 layer is lost, it can undergo corrosion processes, releasing particles that can lead the host into a pro-inflammatory state and tissue damage with a direct effect on cells [8].

Cobalt-chromium (CoCr) alloys are another important option for orthopedic implants. These alloys offer high wear resistance, excellent mechanical properties, and good corrosion resistance, making them ideal for permanent implants subject to movement, such as hip and knee prostheses [9]. However, like stainless steels, they have a high modulus of elasticity compared to bone and present challenges in processing, cost, and potential adverse reactions due to ion release [1].

Biodegradable metallic implants are gaining attention in the biomedical field due to their ability to degrade in the body after fulfilling their function, eliminating the need for a second surgery to remove them [10]. Implants based on magnesium, iron, or zinc are particularly noteworthy. Magnesium is one of the most researched materials for biodegradable implants due to its excellent biocompatibility and suitable mechanical properties. Magnesium alloys degrade in the body, releasing magnesium ions that are absorbed by the organism [11]. The rapid degradation of Mg in the physiological conditions and subsequent loss of mechanical strength is the primary reason for implant failure. In addition, the accumulation of hydrogen gas during the degradation can cause complications, leading to inflammation and pain in the area of implantation [11]. Iron alloys have a slower degradation rate compared to magnesium, which can be beneficial for applications requiring longer support [12]. However, slow degradation can lead to the accumulation of potentially toxic corrosion products [13]. Zinc alloys offer an intermediate degradation rate between magnesium and iron and exhibit good biocompatibility [14].

Despite the significant advantages of metal implants, several challenges remain unsolved. These include (1) preventing the release of metal ions due to corrosion, which can cause toxic reactions and bone resorption, improving long-term biocompatibility, (2) implant bioactivation to provide it with specific biological functions for tissue regeneration, and (3) reducing the risk of infections. This document will focus on the study of how sol-gel silicon coatings can be an excellent alternative to address these challenges and improve the performance of metal implants.

2. Silica sol-gel materials and their potential

Since Larry Hench developed the first bioactive glass for tissue engineering applications in 1969, novel silica materials with improved features, such as those obtained through sol-gel compounds, have emerged, offering new possibilities in the biomedicine field [15]. *In vivo* studies have demonstrated that bioactive glasses bond with bone, and *in vitro* evaluations suggest that their osteogenic properties arise from their dissolution products stimulating osteoprogenitor cells [16]. The application of sol-gel technology to obtain silicon-based biomaterials made it possible to overcome the constraints of traditional bioglasses such as the need of high sintering temperatures or the limited versatility to tuning material characteristics such as porosity, hydrophilicity, or biodegradation kinetics. The sol-gel process involves two types of reactions that can occur simultaneously: hydrolysis and condensation of alkoxysilanes, resulting in a solid silicate network. In the initial step, the alkoxysilane precursor is hydrolyzed by reacting with water to form silanol groups. During condensation, a silanol group interacts with another silanol group or a non-hydrolyzed alkoxy group (-OR), crosslinking the silicon atoms through siloxane bonds (Si-O-Si) and forming silica nanoparticles [17]. Gelation occurs when siloxane links are formed between the dispersed colloidal particles, creating a three-dimensional structure that encapsulates the liquid phase. These materials can be synthesized at low temperatures using heat treatments, obtaining a xerogel [18].

Materials with specific characteristics can be synthesized through the sol-gel process due to the precise control over the chemical composition of the final product. Sol-gel processing allows the use of organically modified alkoxysilanes, enabling the incorporation of functional groups into the sol-gel network to create hybrid compounds [19]. Moreover, this technique presents a high versatility to design biomaterials as the control of synthesis parameters (precursor type, degree of functionalization, temperature, pH, H_2O ratio, and solvent type) allows to create networks with varying attributes in terms of hydrophilicity, morphology, porosity, chemistry, and degradability [20]. On the other hand, these silica sol-gel (or hybrid silica sol-gel) materials have demonstrated their biocompatibility and significant potential in biomedical applications, especially for bone tissue regeneration applications [21]. In fact, silica sol-gel networks have shown promising results for bone regeneration, as they support osteoblast attachment, proliferation, and differentiation, enhancing the expression of multiple osteogenic genes [22].

These materials can be applied as coatings on metallic implants, bioactivating them. The sol-gel structures can usually reach a good adhesion to metal substrates due to the formation of covalent bonds between the sol-gel network and hydroxyl groups present on the metal surface [23]. These coatings can be designed to biodegrade, being able to release silicon compounds in the $Si(OH)_4$ form, which provide osteoinductive properties to the implant, as silicon positively influences bone metabolism by promoting osteoblastogenesis [24]. The silicon compounds released from the sol-gel networks can enhance the differentiation of osteoblasts by promoting the synthesis of type I collagen in human osteoblast-like cells [25]. This type I-collagen, which is a component of the extracellular matrix, was found to boost the expression of mature osteoblastic characteristics, as evidenced by increased alkaline phosphatase activity, osteocalcin synthesis, and mineralization [26]. Moreover, silicon has been reported to reduce osteoclastogenesis and exert a strong inhibitory effect on bone resorption while increasing the rate of bone formation [27].

The sol-gel coatings can be used as delivery systems due to their controllable biodegradability kinetics [28]. Another advantage of the sol-gel technique is that sol-gel reactions can be normally developed at relatively low temperatures, enabling the production of materials in an energy-efficient manner. This approach is particularly useful for incorporating temperature-sensitive drugs or active compounds into the sol-gel network [29]. Therefore, these networks can be enriched with therapeutic agents, adding value to the metallic implants by not only bioactivating them but also serving as a delivery vehicle for elements with specific functions (*e.g.* antibacterial, anti-inflammatory, osteogenic) at the healing site. At the same time, these coatings can be designed to minimize this degradation capacity and serve to protect the metal implant from corrosion, avoiding the release of toxic corrosion products that can compromise the success of the prostheses [30]. In this way, sol-gel technology allows the production of coatings with varied properties that can solve the three main limitations of metal implants. However, an optimal balance between bioactivating and protecting the implant is necessary, being this equilibrium dependent on the characteristics of the metal substrate and the specific objectives of applying the sol-gel coating.

3. Silica sol-gel coatings to improve corrosion resistance

To ensure the success of a metal implant, it is sometimes necessary to protect the material from corrosion, either to prevent the release of toxic compounds or to delay the prosthesis's biodegradation. Researchers have shown that silicon sol-gel coatings can enhance metallic implants' performance by increasing their corrosion resistance. This greater protective effect is a consequence of the physical barrier generated between the metal substrate and the physiological medium in which it is located. Parameters such as the degree of crosslinking of the sol-gel network or the organic load present in its structure are crucial in the effectiveness of the coating in this sense. However, it is important to remember that applying sol-gel materials, which alter the surface properties of the metal, will also change its interaction with the biological environment. Therefore, it is crucial to ensure that these coatings do not negatively impact regeneration processes. In some cases, although not the primary goal, surface modifications will also alter the implant's bioactivity.

Table 1 shows a summary of research addressing the development and characterization of sol-gel Si coatings to prevent corrosion problems in metallic implants.

3.1 Non-biodegradable metallic implants

The durability and long-term success of metallic implants can be enhanced by applying silica sol-gel coatings, which protect the implant surface from corrosion issues and prevent the release of toxic elements from the metallic implant. Various approaches have been carried out by researchers in the field to achieve this goal. Thus, sol-gel coatings manufactured from different precursors have shown good results in this regard. NiTi alloys, with shape memory effects, can show problems due to the release of Ni ions. Zheng et al. [31] developed a $SrO–SiO_2–TiO_2$ sol-gel coating that significantly increased the corrosion resistance of this material and decreased the release of nickel ions. Additionally, this treatment improved attachment, spreading and proliferation of MG63 osteoblasts with respect to the non-coated samples. Aiming to reduce the carcinogenic potential of cobalt-based alloys, a silica-based sol-gel coating was applied on the Co-Cr-Mo substrate [32]. The synthesis of the coatings was based on methyltriethoxysilane (MTES) and tetraethoxysilane (TEOS), resulting in defect-free films on the cobalt alloy. This type of coating showed the ability to protect the metal substrate. In addition, electrochemical tests showed greater corrosion resistance in coatings cured at 450°C compared to others treated at 550°C. The reduction in curing temperature means that the integrity of the organic charge of the sol-gel film is maintained to a greater degree; and the methyl groups present in the coating provided a greater barrier effect. Tiwari et al. [33] developed a sol-gel silica–alumina composite coating to be applied onto SS AISI 316 L for implant applications. The coating was stable and did not degrade after a long time of exposure in Ringer's solution, preventing the leaching of undesired metal ions from the substrate even at high anodic potentials.

Substrate	Presursor	Additive	Curing T (°C)	Coating method	Catalyst	Ref.
NiTi	TEOS, TTIP	$Sr(NO_3)_2$	200	dip-coating	HCl	[31]
Co alloy	TEOS, MTES	—	450/550	dip-coating	Acetic ac	[32]
SS AISI 316 L	TEOS, ACROS	—	500	dip-coating	HNO_3	[33]
Ti6Al4V	DMTMSP, TMOS, MAPTMS	—	120	dip-coating	—	[34]
SS AISI 316 L/ Ti6Al4V	TEOS, GPTMS	TiO_2 NPs	80	dip-coating	HCl	[35]
—	HEMA, TEOS, γMPS, MTES	SiO_2–CaO–P_2O_5	450/150	dip-coating	HNO_3/ acetic ac	[36]
SS AISI 316 L	TEOS, MTMOS, GPTMS	—	140–80	dip-coating	HNO_3	[37]
SS AISI 316 L	TEOS, MTMOS, DMDES, PDMS	—	120	dip-coating	HCl	[38]
Mg alloy	TEOS, $Ce(NO_3)_3$, GMA	—	250	dip-coating	HCl	[39]
Mg alloy	TMOS, MTES, TEPI, Ca L-lactate	—	300	dip-coating	HNO_3	[40]
Mg alloy	TEOS, TEP, $Ca(NO_3)_2{\cdot}4H_2O$	$Mg_3(PO_4)_2$	RT	spin-coating	HCl	[41]
Mg alloy	TTIP, GPTMS	—	120	dip-coating	HNO_3	[42]
Mg alloy	TEOS, GPTMS	SiO_2 NPs	180	dip-coating	HCl/HNO_3	[43]
Mg alloy	APTES, PMMA	—	120	dip-coating	H_2SO_4	[44]
Mg alloy	MTES, GPTMS	—	120	dip-coating	Acetic ac	[45]
Mg alloy	BTSE, γ-APS	heparin	120	dip-coating	—	[46]

Table 1.
Summary of studies based on silica sol-gel coatings for preventing corrosion in metallic implants.

Aiming to protect metallic implants from corrosion, a hybrid sol-gel coating containing 3-(trimethoxysilyl)propyl methacrylate (MAPTMS), tetramethyl orthosilicate (TMOS), and dimethyl trimethylsilylphosphite (DMTMSP) precursors were prepared and applied on Ti6Al4V alloy [34]. The incorporation of DMTMSP into the MAPTMS/TMOS mixture results in the formation of Si-O-P and PO_4^{-3} units, indicating its successful molecular-level integration into the sol-gel network. To improve the implant corrosion resistance, the application of multilayer coatings was necessary. The presence of phosphorus in the matrix enhanced the osteoblast cytotoxicity/viability and adhesion *in vitro*, leading to improved biocompatibility. Łępicka et al. [35] utilized a 3-glycidoxypropyl trimethoxysilane (GPTMS):TEOS coating with TiO_2 nanoparticles to modify the surface of dental implant abutments. Compared to the pure GPTMS:TEOS matrix, the addition of TiO_2 nanoparticles to the sol-gel structure enhanced the coating's resistance to hydrolytic degradation while reducing its hardness and Young's modulus. Ballarre et al. [36] proposed the application of a dual coating based on one TEOS–MTES sol-gel inner layer and a second film obtained through TEOS, hydroxyethyl methacylate (HEMA) and 3-methacryloxypropyl-trimethoxysilane (γMPS) precursors. Additionally, SiO_2–CaO–P_2O_5 particles were incorporated into the sol-gel structure and its effect on coating corrosion resistance was evaluated. The samples with the two types of coatings displayed a protective behavior after 30 days of immersion due to the compact TEOS-MTES inner layer. Additionally, the authors found that particles placed in the outer layer showed a dissolution rate like the rate of deposition of the amorphous hydroxyapatite (HA) at the surface, blocking the entrance of the electrolyte, avoiding it from reaching the substrate. Therefore, the addition of this kind of glass particles to bioactivate the implant does not have to compromise the best performance under corrosion conditions. In another work, sol-gel hybrid coatings obtained from the combination of three different alkoxysilanes (methyltrimethoxysilane (MTMOS), GPTMS, and TEOS) were applied on AISI316-L stainless steel [37]. The coatings showed an improvement in the anticorrosive properties of the steel. Additionally, degradation and electrochemical impedance spectroscopy results showed that by controlling the amount of TEOS precursor in the coating it is possible to tune its degradation by hydrolysis, while keeping properties such as wettability at their optimum values for biomaterials application. Sol-gel coatings made using methyl-modified alkoxysilanes were developed to bioactivate AISI 316 SS [38]. Three precursors functionalized with methyl groups in different configurations were evaluated (MTMOS, polydimethylsiloxane (PDMS), and dimethyldiethoxysilane (DMDES)) and mixed with TEOS (non-functionalized). In all cases, homogeneous, well-adhering coatings were obtained on the metallic substrate. The corrosion protection of the coatings and their hydrolytic degradation varied with the precursor used. MTMOS-based coatings released the most Si compounds, while PDMS and DMDES provided better protection. DMDES-based coating offered an interesting balance between protection and bioactivation.

3.2 Biodegradable metallic implants

As for biodegradable metal implants, sol-gel coatings are used to control their corrosion speed and to improve their corrosion resistance in the physiological environment. Reyes et al. [39] modified the surface of Mg alloy by the deposition of sol-gel Ce_xO_y/SiO_2 coatings, as the presence of cerium compounds acts as inhibitors activating corrosion protection through a "self-healing" behavior. These coatings decreased the degradation rate of the Mg alloy and increased its bioactivity. Omar et al. [40] retarded the AZ91D Mg alloy degradation by applying bioactive sol-gel glass coatings. Additionally, the developed treatments promoted the adhesion and proliferation of pre-myoblast auto-fluorescent cells. Another approach was the application of magnesium phosphate/glass composite materials as a coating on Mg alloy substrates [41]. These composites were based on a SiO_2-CaO-5P_2O_5 sol-gel network with $Mg_3(PO_4)_2$ particles dispersed in the glass structure. These magnesium phosphate/glass composite coatings enhanced the formation of an apatite-bone-like layer on the Mg metal surface, and they were viable with oral epithelial cells. Additionally, the corrosion resistance was improved for coatings containing up to 10% wt of Mg-phosphate particles. TiO_2-modified silane coatings were developed to gradually regulate the corrosion rate of Mg alloys [42]. The sol-gel treatments, which were made using titanium tetraisopropoxide (TTIP) and GPTMS, demonstrated an immediate barrier effect. This fact can be related to the significant reduction in cracks, pits, and other structural defects on the surfaces of the coated alloys. Additionally, the silane coatings promoted the formation of calcium phosphate compounds, indicating increased bioactivity. In another study, Castro and Duran [43] developed a sol-gel coating based on TEOS and GPTMS precursors with SiO_2 nanoparticles. This material was applied in Mg alloys, demonstrating their capacity to block the degradation of the alloy during 8 days in simulated body fluid. The electrochemical corrosion tests displayed a relevant effect on the protection of Mg alloys by the deposition of this silica film. A composite coating based on (3-Aminopropyl)triethoxysilane (APTES) and polymethylmethacrylate (PMMA) was applied on an Mg alloy to improve its corrosion behavior [44]. The enhanced corrosion resistance of the composite coating was primarily attributed to the superior morphological characteristics of the sol-gel film, including their uniformity and crack-free surface. Additionally, this improvement was related to the amide bond formed between the amine group of APTES and the ester of PMMA. Gaur et al. [45] developed a GPTMS:MTES sol-gel coating to be applied on a Mg alloy, which demonstrated significant protective properties. The coating notably reduced hydrogen evolution and minimized the increase in pH, both desirable outcomes for the use of Mg alloys as implants. *In vitro* cell culture results clearly showed the superiority of the silane-coated Mg alloy with respect to the non-coated samples. Lie et al. [46] developed a two-step sol-gel coating process as a strategy to create a biofunctionalized, anti-corrosive layer on the Mg alloy AZ31, intended for use in cardiovascular stents. They used bistriethoxysilylethane (BTSE) and 3-amino-propyltrimethoxysilane (γ-APS) as precursors to treat the implant surface, followed by functionalization of the coating with heparin. Compared to the untreated AZ31 Mg alloy, the surface treatments demonstrated both enhanced corrosion resistance and reduced platelet adhesion. This surface modification strategy not only improves the corrosion resistance of the Mg alloy but also inhibits platelet adhesion, offering significant potential for enhancing the blood compatibility of biodegradable metallic implants.

4. Silica sol-gel coatings to bioactivate metallic implants

As previously mentioned, sol-gel silicon coatings present a promising approach to enhance the tissue regeneration potential of bioinert metal implants. This bioactivation can be achieved directly by modifying the implant's surface with the sol-gel coating. Additionally, specific functionalities can be introduced by incorporating active agents into the sol-gel matrix, which then acts as a release vehicle. Several researchers have successfully addressed this method of bioactivating metal implants, particularly for bone regeneration applications, as demonstrated by the studies summarized in **Table 2**.

Silica/polyethylene glycol (PEG) hybrid coatings were applied onto Ti grade 4 substrates to improve their biological properties [47]. These coatings were more bioactive and biocompatible than the uncoated Ti surfaces to NIH 3 T3 fibroblasts. Additionally, the results showed that the PEG amount incorporated into the silica structure did not significantly affect the coating bioactivity. Hybrid coatings made of a silica xerogel/chitosan matrix were applied to Ti [48]. Increasing the silica xerogel content in these hybrid coatings enhanced their *in vitro* bioactivity, as evidenced by improved cell attachment and alkaline phosphatase (ALP) activity in MC3T3-E1 cells. Ballarre et al. [49] developed a surface treatment using a double sol-gel MTES-TEOS coating to enhance the osseointegration capacity of stainless steel implants. The inner layer contained SiO_2 colloidal particles, while the outer layer incorporated SiO_2–P_2O_5–CaO ceramic particles. The SiO_2 colloidal particles increased the barrier effect, preventing the diffusion of potentially toxic ions. This sol-gel system improved the bioactivity *in vitro* and promoted the formation of new bone around the implant, both in the remodeling zone and the marrow zone.

Substrate	Presursor	Additive	Curing T (°C)	Coating method	Catalyst	Ref.
Ti	TEOS, PEG	—	45	dip-coating	HNO_3	[47]
Ti	TMOS, Chitosan	Triethyl phosphate	37	spin-coating	Acetic ac	[48]
SS AISI 316 L	TEOS, MTES	SiO_2 particles/ SiO_2–P_2O_5–CaO	450	dip-coating	HNO_3	[49]
Ti	TEOS, PCL	—	45	dip-coating	HNO_3	[50]
SS AISI 316 L	TEOS, ZrOP	—	250	dip-coating	HCl	[51]
	TEOS, VTES	—	100	drop-coating	HNO_3	[52]
SS AISI 316 L, Ti	TEOS, GPTMS	—	140	dip-coating	HNO_3	[53, 55]
Ti	TEOS, MTMOS	—	80	dip-coating	HNO_3	[56, 57]
Ti	TITP, TEOS	—	40	dip-coating	HCl, HNO_3	[58, 59]
Mg alloy	TEOS, MTES	Graphene NPs	200	dip-coating	HCl	[60]
Ti	TEOS, MTMOS, GPTMS	$SrCl_2$	80	dip-coating	HNO_3	[61]
Ti	TEOS, MTMOS	$SrCl_2$	80	dip-coating	HNO_3	[62]
SS AISI 316 L	TEOS, MTES	BG,Sr-BG particles	450	dip-coating	—	[63]
SS AISI 316 L	TEOS, MTES	Wollastonite	450	dip-coating	HNO_3	[64]
Ti	TEOS, MTMOS	$ZnCl_2$	80	dip-coating	HNO_3	[65]
Ti	TEOS, MTMOS	$MgCl_2$	80	dip-coating	HNO_3	[66]
Ti	TEOS, MTMOS, GPTMS	$CaCl_2$	80	dip-coating	HNO_3	[67]
Ti	TEOS, MTMOS	$CaCl_2$	80	dip-coating	HNO_3	[68]
Ti	TEOS, MTMOS	$CaCl_2$, $MgCl_2$	80	dip-coating	HNO_3	[69]
Ti	TEOS, MTMOS	$CaCl_2$, $ZnCl_2$	80	dip-coating	HNO_3	[70]
SS AISI 316 L	TEOS, MTMOS/ GPTMS, MTMOS	Gelatin	80	dip-coating	HCl	[71]
Ti	TEOS, MTMOS/ GPTMS, MTMOS, TEOS	Gelatin	80	dip-coating	HNO_3	[72]
Ti	TEOS, MTMOS	Melatonin	80	dip-coating	HCl	[73]

Table 2.
Summary of studies based on silica sol-gel coatings to bioactivate metallic implants.

Hybrid sol-gel coatings based on TEOS and poly (e-caprolactone) (PCL) were obtained to bioactivate Ti implants [50]. PCL is a biocompatible aliphatic polyester that has been shown to act as a plasticizer, giving elasticity to the sol-gel matrix and enabling to obtain crack-free coatings. This SiO_2/PCL coating improved the NIH 3 T3 fibroblast *in vitro* biological performances of the grade-4 Ti. However, the better cellular response was mainly provided by the silica inorganic component of this hybrid matrix. Smieszek et al. [51] developed a SiO_2/ZrO_2 sol-gel coating to bioactivate 316 L

SS implants. They compared the *in vitro* biological outcome of that SiO_2/ZrO_2 system with respect to a ZrO_2 sol-gel coating in mesenchymal stem cells (MSCs). The proliferation of MSCs was accelerated with the SiO_2/ZrO_2 coating compared to the ZrO_2. Moreover, the SiO_2/ZrO_2 material exhibited greater osteoinductive properties with enhanced expression of ALP and osteopontin (OPN). Hernández-Escolano et al. [52] developed hybrid sol-gel materials based on vinyl tri-ethoxysilane (VTES) and TEOS to bioactivate metallic implants. This sol-gel system displayed osteogenic properties, being the MSCs mineralization capacity stimulated by increasing the TEOS content in the coating. Juan-Diaz et al. [53] synthetized MTMOS:GPTMS coatings aiming to promote dental implant osseointegration. The increase of the GPTMS content raised the wettability and water absorption of the coating. Moreover, the addition of GPTMS to MTMOS significantly increased the release of Si compounds, which was accompanied by enhanced *in vitro* proliferation and mineralization. Despite these promising results, it is essential to carefully adjust the composition of the GPTMS precursor in sol-gel formulations intended for tissue regeneration applications. Although GPTMS is non-cytotoxic, and the epoxy ring in its structure facilitates the incorporation of osteogenic or antibacterial drugs into the sol-gel network by covalent bonding; its characteristic functional group is susceptible to nucleophilic attack [54]. Therefore, sol-gel coatings with GPTM tend to promote the adsorption of immune proteins, and in parallel, favored the pro-inflammatory immune response *in vitro*. In fact, sol-gel coatings with more than 35% GPTMS were associated with a polarization of M1-phenotype macrophages and the formation of fibrous connective tissue *in vivo*, evidencing biocompatibility problems [55]. Martínez-Ibáñez et al. [56] combined different amounts of MTMOS and TEOS to obtain a coating with which to bioactivate Ti dental implants. *In vitro* studies with human adipose tissue-derived mesenchymal stem cells showed the highest levels of proliferation and mineralization for the coating with 7:3 MTMOS:TEOS ratio. *In vivo* studies on dental implants coated with this sol-gel material indicated that the released Si enhanced bone regeneration, as evidenced by the earlier formation of new bone spicules. Additionally, the study of the regeneration mechanism around this 7MTMOS:3TEOS coated dental implants was proved to be different from that observed for untreated screws, being the degree of osseointegration, in the case of the sol-gel coating, mainly achieved by osteoinduction as a result of the release of Si compounds [57]. Ääritalo et al. [58] developed a series of sol-gel derived TiO_2–SiO_2 mixed oxide coatings to be applied on Ti implants and favor bone tissue regeneration. These titania–silica mixed coatings were made using tetraisopropylorthotitanate (TIPT) and TEOS as precursors. A two-step hydrolysis of the Ti-alkoxide and Si-alkoxide solutions was made, resulting in heterogeneously mixed oxide phases. The coatings demonstrated an improvement in the fibroblast proliferation, with prolonged osteoblast activity observed as the amount of released silica increased. Additionally, after 21 days of culture with osteoblasts, the surfaces were completely covered with calcified nodules and extracellular matrix [59].

5. The use of the silica sol-gel coating as release vehicles to bioactivate the metallic implant

The following sections will discuss advances in the use of silicon sol-gel coatings as vehicles for the release of bioactive agents. Sol-gel networks have been employed to deliver both inorganic and organic compounds, which have shown promising effects on tissue regeneration around metallic prostheses. Additionally, the role of sol-gel coatings in preventing infections by incorporating antimicrobial agents into their structure is particularly noteworthy. Bioactive agents can sometimes be directly incorporated into the sol-gel network, where they are either physically trapped or covalently bonded to functional groups within the structure. In other cases, to slow their release, these compounds are first embedded in carriers, such as nanoparticles, which are then incorporated into the sol-gel matrix.

5.1 Inorganic agents

Fernández-Hernán et al. [60] improved the biological performance of the AZ31 alloy through its coating with a nanocomposite based on a silica sol-gel material functionalized with graphene nanoplatelets (GNPs). The addition of 0.05 wt% GNPs significantly delayed the beginning of the Mg alloy degradation process and improved MC3T3 pre-osteoblastic cells cytocompatibility and adhesion *in vitro*. Aiming to increase the bone regenerative potential of a hybrid sol-gel coating obtained through MTMOS, TEOS, and GPTMS precursors, $SrCl_2$ was incorporated into its structure [61]. Sr. was reported to stimulate bone growth and reduce its resorption, showing a positive effect on osteogenesis. Its incorporation into the sol-gel coating confers greater osteogenic properties to the material, with increased ALP and transforming growth factor (TGF)-β expression *in vitro*. In this regard, García-Arnáez et al. [62] developed a Sr-enriched sol-gel coating from TEOS and MTMOS alkoxysilanes to bioactivate Ti dental implants. Sr. doping not only augmented the coating osteogenic potential but also improved its anti-inflammatory response *in vitro*. Omar et al. [63] evaluated the application of a double layer of sol-gel coatings functionalized either with silica particles (BG) or Sr-substituted BG (Sr-BG) particles onto surgical-grade stainless steel. These coatings were able to induce the formation of HA on their surfaces immersed in simulated body fluid (SBF), and enhanced corrosion resistance when compared with bare SS. In addition, the coating containing Sr-BG particles showed better bone regeneration at early times of implantation *in vivo* in Wistar–Hokkaido male adult rats. MTES:TEOS coating system incorporating wollastonite particles produced homogeneous coatings on surgical stainless steel and improved their corrosion resistance [64]. This silica sol-gel/wollastonite coating promoted new bone formation. In addition, nanoindentation studies revealed that the new bone regenerated around the coating had good nanomechanical properties, comparable to remodeled cortical bone, though it showed lower modulus and greater brittleness compared to pre-existing bone. Cerqueira et al. [65] developed Zn-doped sol-gel coatings using MTMOS and TEOS as precursors and incorporating different amounts of $ZnCl_2$ during the synthesis of the sol-gel material. The sol-gel structure allowed a controlled Zn^{2+} release which ensured the non-cytotoxicity of the films. The addition of Zn increased ALP, TGF-β, and RUNX2 (runt-related transcription factor 2) expression in osteoblasts, which revealed a higher osteogenic potential with respect to the non-doped sol-gel coating. Moreover, the immune response of these materials was evaluated *in vitro* with macrophages, showing an increased expression of TGF-β and interleukin (IL)-4 anti-inflammatory markers, accompanied by a reduced tumor necrosis factor (TNF)-α (a pro-inflammatory cytokine). In another study, hybrid silica sol-gel coatings able to release Mg^{2+} were developed to improve cell adhesion and osteogenic potential of Ti dental implants [66]. The cytoskeleton organization of MC3T3-E1 cells cultured on the Mg-doped sol-gel coatings exhibited a more elongated shape with extended lamellipodia and an enlarged cell surface with respect to the controls. The expression of integrins (ITGA5 and ITGB1) was enhanced in osteoblasts cultured on these Mg-coatings. Additionally, there was an increase in ALP activity and the expression of osteoblast-specific transcription factor osterix (OSX), TGF-β, and RUNX2 osteogenic genes.

Calcium is widely used in biomaterials for bone tissue engineering due to its key roles in bone metabolism and blood coagulation. Therefore, some authors assessed the incorporation of this element into silica sol-gel structures to enhance the regenerative potential of this kind of biomaterial. Incorporating $CaCl_2$ into an MTMOS-GPTMS-TEOS sol-gel network caused a rapid release of Ca^{2+} within the first 24 hours, followed by a steady release over a month [67]. This calcium addition enhanced *in vitro* osteogenic potential compared to coatings without it. Additionally, the presence of Ca in the sol-gel coating significantly increased blood coagulation activity on implant surfaces by altering protein adsorption patterns. Ca-doped hybrid sol-gel coatings applied on Ti displayed a dose-dependent pro-inflammatory response *in vitro* [68]. *In vivo*, in a rabbit condyle model 5 days after surgery, screws coated with these Ca-sol-gel materials showed the formation of immune structures and an increment of erythrocyte clumps for the sample with the highest amount of $CaCl_2$ (5%wt). Sol-gel coatings based on MTMOS and TEOS enriched with combinations of bioactive cations were also studied. Thus, $CaCl_2$ with a strong osteogenic and coagulatory effect was combined with ions with anti-inflammatory potential such as Mg^{2+} [69] or Zn^{2+} [70] giving rise to sol-gel coatings able to release various elements capable of promoting bone regeneration.

5.2 Organic agents

Taking into account that natural polymers, such as gelatin, can mimic collagen I and enhance the implant biocompatibility, Martínez-Ibáñez et al. [71] assessed the immobilization strategies and dissolution properties of gelatin in different silica sol-gel matrices to create siloxane-gelatin hybrid coatings for enhancing the bioactivation of metallic implants. They found that the GPTMS precursor can effectively serve as a covalent coupling agent between the silica and gelatin phases, increasing the stability of the coatings in aqueous media. In another study, two sol-gel materials doped with 0.9 wt% gelatin were biologically tested when applied as coatings on Ti surfaces [72]. One of the sol-gel networks was based on a combination of MTMOS and TEOS precursors, while the other one had GPTMS in its composition in addition to MTMOS and TEOS, which allowed covalent immobilization of gelatin to the sol-gel structure. The incorporation of gelatin into the sol-gel improved proliferation and ALP activity in MC3T3-E1. However, although both sol-gel coatings were able to support new bone regeneration *in vivo*, an increased inflammatory reaction was observed by the greater presence of multinucleated giant cells around both gelatin-supplemented coatings. Additionally, the incorporation of gelatin into the sol-gel coatings increased the adsorption of C reactive protein on the implants, which is associated with a higher immune response. MTMOS-TEOS sol-gel coatings doped with melatonin were developed to promote the osseointegration of Ti dental implants [73]. Melatonin is a widely known hormone able to regulate the circadian cycles and is studied for its role in bone regeneration and inflammation. The studied sol-gel films increased gene expression of RUNX2 and bone morphogenic protein (BMP)-2 in MC3T3-E1 cells, as well as decreased the production of TNF-α in RAW 264.7 macrophages, showing an anti-inflammatory potential.

5.3 Antimicrobial agents

The potential of silica sol-gel coatings to release antimicrobial compounds preventing the risk of infections in metallic implants has been evaluated by several authors. **Table 3** shows a summary of the main advances in this regard.

Radin et al. [74] applied TEOS-based sol-gel coatings on a Ti6Al4V substrate and loaded them with vancomycin. They studied the effect of the processing parameters on the coating degradation and antibiotic release, reporting the long-term, controlled release of vancomycin from these films. MAPTMS:TMOS:tris (trimethylsilyl)phosphite (TMSPi) organic–inorganic sol-gel coatings were tested as carries of seven different antibiotics (gentamicin, cefoxitin, vancomycin, dicloxacillin, clindamycin, ampicillin and, linezolid) and the antifungal amphotericin B [75]. The introduction of these antimicrobial molecules into the sol-gel structure affected the network crosslinking but also roughness, thickness, and wettability of the resulted coatings. This MAPTMS:TMOS:TMSPi system loaded with linezolid and cefoxitin antibiotics was applied as coating to functionalize Ti implants [76]. Although the MAPTMS:TMOS:TMSPi sol-gel coating maintained its integrity for the first 24 h (in terms of degradability), it allowed sufficient release of linezolid to inhibit the *Staphylococcus aureus* and *Staphylococcus epidermidis* biofilm formation. On the other hand, cefoxitin release was compromised by its chemically bonding during the sol-gel synthesis; fact that diminished the antibacterial effect of the coatings with this antibiotic. Anyway, loading both antibiotics in the sol-gel network showed synergistic effects in its antibacterial potential.

Biodegradable coatings for the controlled release of antifungals in Ti joint prostheses were developed using a MAPTMS-TMOS sol-gel system [77]. The sol-gel networks loaded with anidulafungin displayed greater degradation kinetics than those with fluconazole. The antifungal agent release depends on the degradation of the coatings. The results revealed that anidulafungin-doped coatings were more effective in inhibiting the formation of *Candida albicans* biofilm than those with fluconazole, showing a promising feature to locally prevent prosthetic joint infections. Hybrid sol-gel coatings enriched with $CuCl_2$ were synthesized from MTMOS and TEOS and applied on Ti grade 4 for dental applications [78]. Human osteoblasts in contact with these materials exhibited increased osteogenic potential with higher ALP, BMP2, and osteocalcin (OCN) gene expressions with respect to the coating without Cu. In turn, the increase in the amount of $CuCl_2$ in the sol-gel network led to an augmented pro-inflammatory response of the coatings *in vitro*. Additionally, the copper-doped coatings showed strong antibacterial activity against *Escherichia coli* and *S. aureus*. A double coating system to protect SS implants and confer them antibacterial properties was developed using TEOS and MTMOS as precursors [79]. In this work, three kinds of BG particles (silicate, borosilicate, and Cu-doped borosilicate) were incorporated into the sol-gel outer layer. The borosilicate and Cu-doped borosilicate particles were able to inhibit the *E. coli* growth. Additionally, the studied coatings presented excellent anti-corrosive behavior and bioactivity (measured by the HA deposition). However, the antibacterial potential of the doped coatings themselves was not evaluated. Molina et al. [80] gave antibacterial properties to Ti6Al4V alloy by using a GPTMS:TEOS silica-based sol-gel coating with Ag-NPs. An antibacterial hydroxyapatite coating with Ag NPs was also applied to compare the performance of both coatings. The antibacterial activity of both treatments was tested against *S. aureus* and *S. epidermidis*, demonstrating a 99.9% lethality compared to the sol-gel coatings without Ag-NPs. However, the silica-based coating had the benefit of

Substrate	Presursor	Additive	Curing T (°C)	Coating method	Catalyst	Ref.
Ti6Al4V	TEOS	Vancomycin	RT	dip-coating	HCl	[74]
Ti	MAPTMS, TMOS, TMSPi	Gentamicin, cefoxitin, vancomycin, dicloxacillin, clindamycin, ampicillin, linezolid, amphotericin B	60	dip-coating	—	[75]
Ti	MAPTMS, TMOS, TMSPi	Linezolid, cefoxitin	60	dip-coating	—	[76]
Ti	MAPTMS, TMOS	Fluconazole, anidulafungin	60	dip-coating	—	[77]
Ti	TEOS, MTMOS	$CuCl_2$	80	dip-coating	HNO_3	[78]
SS AISI 316 L	TEOS, MTMOS	BG, B-BG, Cu-B-BG particles	450	dip-coating	HNO_3	[79]
Ti6Al4V	TEOS, GPTMS	$AgNO_3$	100	dip-coating	HNO_3	[80]
Ti	TEOS, MTMOS, GPTMS	Chitosan	100	dip-coating	HCl	[81]
Ti	TEOS, MTES/gelatin, chitosane	BG particles/Si-gentamicin NPs	450/RT	spray/electrophoretic deposition	HNO_3/Acetic ac	[82]
Ti	TEOS, MTMOS	Chlorhexidine diacetate	80	dip-coating	HNO_3	[83]

Table 3.
Summary of studies based on silica sol-gel coatings to confer antimicrobial properties to metallic implants.

low-temperature synthesis, as well as excellent Ag-NP distribution with a sustained release of the antibacterial compound for up to 40 days.

Chitosan has antimicrobial activity by its positively charged amino groups which can interact with the negatively charged bacterial cell wall, killing them. Considering the effect of this compound, Palla-Rubio et al. [81] developed a silica sol-gel matrix containing chitosan to be applied on Ti dental implants. The coatings with 5–10% chitosan in their structure displayed substantial antibacterial properties against *S. aureus* strain. Another approach to impart antibacterial properties to Ti implants involved a dual coating system [82]. The first layer consisted of a sprayed MTES:TEOS hybrid sol-gel network with BG particles, while the outer layer was composed of a chitosan-gelatin matrix with silica-gentamicin NPs, applied via electrophoretic deposition. This double-coating system demonstrated effective antibacterial activity against *S. aureus* and *E. coli*. García-Arnáez et al. [83] described the obtention of hybrid sol-gel coatings with chlorhexidine diacetate to be applied on metallic bone implants. These materials displayed a strong bactericidal effect, resulting non-cytotoxic for human osteoblast.

6. Conclusions

In this chapter, the main research challenges for enhancing the performance of metal implants were identified. It was shown that Si sol-gel coatings offer a promising approach to functionalizing these implants, thereby improving their functionality and extending their long-term life. Sol-gel technology enables the surface modification of metal implants in a straightforward and cost-effective manner and was proven to produce uniform coatings that adhere well to metal substrates. These coatings are typically derived from sol-gel compositions based on organo-modified alkoxysilanes and acid catalysis, with the sol-gel mixture often applied via dip-coating.

The studies here exposed confirmed that these materials can significantly improve the corrosion resistance of metal implants, ensuring their biocompatibility by preventing the release of toxic corrosion products. Additionally, silicon sol-gel coatings can help to regulate the degradation rates of biodegradable metal implants, ensuring their proper function over time. These coatings also enable the bioactivation of metallic prostheses, showing considerable potential in implants designed for contact with bone tissue. Both *in vitro* and *in vivo* results have demonstrated that these sol-gel coatings can enhance the biological response of metallic implants. Moreover, the potential of these sol-gel matrices as vehicles for the controlled release of organic and inorganic bioactive agents is noteworthy. The sol-gel networks allow for the adjustment of the release kinetics of bioactive compounds. This versatility has enabled the development of surface treatments that impart antimicrobial properties to metal implants without compromising their biocompatibility, which is crucial for preventing infections.

Looking ahead, given the promising results achieved in the studies presented, it is essential to demonstrate the efficacy of silicon sol-gel coatings in clinical trials. This is a critical step toward bringing this type of treatment to the market in the future.

Acknowledgements

This work was supported by the Science and Innovation Ministry of Spain [PID2023-150429OB-I00 MCIU/AEI/10.13039/501100011033/FEDER/EU].

The authors would like to thank the members of the PIMA research group at the Jaume I University and to researchers Isabel Goñi and Mariló Gurruchaga from the Department of Polymers and Advanced Materials: Physics, Chemistry and Technology at the University of the Basque Country, who have been researching the development of silicon sol-gel materials for biomedical applications for over 10 years.

Conflict of interest

The authors declare no conflict of interest.

Author details

Francisco Romero-Gavilán* and Julio Suay
Department of Industrial Systems Engineering and Design, Universitat Jaume I, Castellón de la Plana, Spain

*Address all correspondence to: gavilan@uji.es

References

[1] Bandyopadhyay A, Mitra I, Goodman SB, Kumar M, Bose S. Improving biocompatibility for next generation of metallic implants. Progress in Materials Science. 2023;**133**:101053. DOI: 10.1016/j.pmatsci.2022.101053

[2] Sultana N, Nishina Y, Nizami MZI. Surface modifications of medical grade stainless steel. Coatings. 2024;**14**(3):248. DOI: 10.3390/coatings14030248

[3] Galliano P, De Damborenea JJ, Pascual MJ, Durán A. Sol-gel coatings on 316L steel for clinical applications. Journal of Sol-Gel Science and Technology. 1998;**13**:723-727. DOI: 10.1023/a:1008653208083

[4] Bekmurzayeva A, Duncanson WJ, Azevedo HS, Kanayeva D. Surface modification of stainless steel for biomedical applications: Revisiting a century-old material. Materials Science and Engineering: C. 2018;**93**:1073-1089. DOI: 10.1016/j.msec.2018.08.049

[5] Williams DF. Titanium for medical applications. In: Brunette DM, Tengvall P, Textor M, Thomsen P, editors. Titanium in Medicine: Material Science, Surface Science, Engineering, Biological Responses and Medical Application. Berlin: Springer; 2001. pp. 12-24

[6] Oliver J, Su Y, Lu X, Kuo PH, Du J, Zhu D. Bioactive glass coatings on metallic implants for biomedical applications. Bioactive Materials. 2019;**4**:261-270. DOI: 10.1016/j.bioactmat.2019.09.002

[7] Kaur M, Singh K. Review on titanium and titanium based alloys as biomaterials for orthopaedic applications. Materials Science and Engineering: C. 2019;**102**:844-862. DOI: 10.1016/j.msec.2019.04.064

[8] Kandaswamy E, Harsha M, Joshi VM. Titanium corrosion products from dental implants and their effect on cells and cytokine release: A review. Journal of Trace Elements in Medicine and Biology. 2024;**84**:127464. DOI: 10.1016/j.jtemb.2024.127464

[9] Mani G, Porter D, Collins S, Schatz T, Ornberg A, Shulfer R. A review on manufacturing processes of cobalt-chromium alloy implants and its impact on corrosion resistance and biocompatibility. Journal of Biomedical Materials Research Part B: Applied Biomaterials. 2024;**112**(6):e35431. DOI: 10.1002/jbm.b.35431

[10] Prakasam M, Locs J, Salma-Ancane K, Loca D, Largeteau A, Berzina-Cimdina L. Biodegradable materials and metallic implants-A review. Journal of Functional Biomaterials. 2017;**8**(4):44. DOI: 10.3390/jfb8040044

[11] Riaz U, Shabib I, Haider W. The current trends of Mg alloys in biomedical applications—A review. Journal of Biomedical Materials Research Part B: Applied Biomaterials. 2019;**107**(6):1970-1996. DOI: 10.1002/jbm.b.34290

[12] Hermawan H, Alamdari H, Mantovani D, Dubé D. Iron–manganese: New class of metallic degradable biomaterials prepared by powder metallurgy. Powder Metallurgy. 2008;**51**(1):38-45. DOI: 10.1179/174329008X284868

[13] Li H, Zheng Y, Qin L. Progress of biodegradable metals. Progress in Natural Science: Materials International. 2014;**24**(5):414-422. DOI: 10.1016/j.pnsc.2014.08.014

[14] Liu Q, Li A, Liu S, Fu Q, Xu Y, Dai J, et al. Cytotoxicity of biodegradable zinc

and its alloys: A systematic review. Journal of Functional Biomaterials. 2023;**14**(4):206. DOI: 10.3390/jfb14040206

[15] Jones JR. Review of bioactive glass: From Hench to hybrids. Acta Biomaterialia. 2015;**23**(S):S53-S82. DOI: 10.1016/j.actbio.2015.07.019

[16] Renno ACM, Bossini PS, Crovace MC, Rodrigues ACM, Zanotto ED, Parizotto NA. Characterization and in vivo biological performance of biosilicate. BioMed Research International. 2013;**2013**:141427. DOI: 10.1155/2013/141427

[17] Reuter H. Sol-gel processes. Advanced Materials. 1991;**3**(5):258-259. DOI: 10.1002/adma.19910030510

[18] Brinker CJ, Scherer GW. Sol-Gel Science - The Physics and Chemistry of Sol-Gel Processing. San Diego: Elsevier Academic Press; 1990

[19] Barrino F. Hybrid organic–inorganic materials prepared by sol–gel and sol–gel-coating method for biomedical use: Study and synthetic review of synthesis and properties. Coatings. 2024;**14**:425. DOI: 10.3390/coatings14040425

[20] Owens GJ, Singh RK, Foroutan F, Alqaysi M, Han CM, Mahapatra C, et al. Sol-gel based materials for biomedical applications. Progress in Materials Science. 2016;**77**:1-79. DOI: 10.1016/j.pmatsci.2015.12.001

[21] Tsuru K, Robertson Z, Annaz B, Gibson IR, Best SM, Shirosaki Y, et al. Sol-gel synthesis and in vitro cell compatibility analysis of silicate-containing biodegradable hybrid gels. Key Engineering Materials. 2008;**361-363**:447-450. DOI: 10.4028/www.scientific.net/KEM.361-363.447

[22] Arcos D, Vallet-Regí M. Sol-gel silica-based biomaterials and bone tissue regeneration. Acta Biomaterialia. 2010;**6**(8):2874-2888. DOI: 10.1016/j.actbio.2010.02.012

[23] Brinker CJ, Scherer GW. Sol-Gel Science: The Physics and Chemistry of Sol-Gel Processing. Boston: Academic Press Inc.; 2013. 908 p. DOI: 10.1016/C2009-0-22386-5

[24] Kim EJ, Bu SY, Sung MK, Choi MK. Effects of silicon on osteoblast activity and bone mineralization of MC3T3-E1 cells. Biological Trace Element Research. 2013;**152**(1):105-112. DOI: 10.1007/s12011-012-9593-4

[25] Reffitt DM, Ogston N, Jugdaohsingh R, Cheung HFJ, Evans BAJ, Thompson RPH, et al. Orthosilicic acid stimulates collagen type 1 synthesis and osteoblastic differentiation in human osteoblast-like cells in vitro. Bone. 2003;**32**(2):127-135. DOI: 10.1016/s8756-3282(02)00950-x

[26] Ghanaati SM, Thimm BW, Unger RE, Orth C, Kohler T, Barbeck M, et al. Collagen-embedded hydroxylapatite-beta-tricalcium phosphate-silicon dioxide bone substitute granules assist rapid vascularization and promote cell growth. Biomedical Materials. 2010;**5**(2):25004. DOI: 10.1088/1748-6041/5/2/025004

[27] Rico H, Gallego-Lago JL, Hernandez ER. Effect of silicon supplement on osteopenia induced by ovariectomy in rats. Calcified Tissue International. 2000;**66**(1):53-55. DOI: 10.1007/s002230050010

[28] Juan-Díaz MJ, Martínez-Ibáñez M, Hernández-Escolano M, Cabedo L, Izquierdo R, Suay J, et al. Study of the degradation of hybrid sol-gel coatings in aqueous medium. Progress in Organic

Coatings. 2014;**77**(11):1799-1806. DOI: 10.1016/j.porgcoat.2014.06.004

[29] Tshikovhi A, Koao LF, Malevu TD, Linganiso EC, Motaung TE. Dopants concentration on the properties of various host materials by sol-gel method: Critical review. Results in Materials. 2023;**19**:100447. DOI: 10.1016/j.rinma.2023.100447

[30] Ghaffari M, Barzegar A, Janghorban K, Javidi M. Investigation of effective parameters in improving corrosion resistance of silica coated stainless steel via sol-gel method. Corrosion Engineering, Science and Technology. 2011;**46**(5):605-610. DOI: 10.1179/147842210X12680457338095

[31] Zheng CY, Nie FL, Zheng YF, Cheng Y, Wei SC, Ruan L, et al. Enhanced corrosion resistance and cellular behavior of ultrafine-grained biomedical NiTi alloy with a novel SrO-SiO_2 -TiO_2 sol-gel coating. Applied Surface Science. 2011;**257**(13):5913-5918. DOI: 10.1016/j.apsusc.2011.02.006

[32] Amato LE, López DA, Galliano PG, Ceré SM. Electrochemical characterization of sol-gel hybrid coatings in cobalt-based alloys for orthopaedic implants. Materials Letters. 2005;**59**(16):2026-2031. DOI: 10.1016/j.matlet.2005.02.010

[33] Tiwari SK, Mishra T, Gunjan MK, Bhattacharyya AS, Singh TB, Singh R. Development and characterization of sol-gel silica-alumina composite coatings on AISI 316L for implant applications. Surface and Coatings Technology. 2007;**201**(16-17):7582-7588. DOI: 10.1016/j.surfcoat.2007.02.026

[34] El hadad AA, García-Galván FR, Mezour MA, Hickman GJ, Soliman IE, Jiménez-Morales A, et al. Organic-inorganic hybrid coatings containing phosphorus precursors prepared by sol–gel on Ti6Al4V alloy: Electrochemical and in-vitro biocompatibility evaluation. Progress in Organic Coatings. 2020;**148**:105834. DOI: 10.1016/j.porgcoat.2020.105834

[35] Łępicka M, Barros-Silva S, Licciardello N, Cortez A, Gobbo P, Sampaio M, et al. Silane-based coating charged with TiO_2 NPs for dental implant applications. Surface and Coatings Technology. 2021;**413**:127066. DOI: 10.1016/j.surfcoat.2021.127066

[36] Ballarre J, López D, Rosero NC, Durán A, Aparicio M, Ceré SM. Electrochemical evaluation of multilayer silica-metacrylate hybrid sol-gel coatings containing bioactive particles on surgical grade stainless steel. Surface and Coatings Technology. 2008;**203**(1-2):80-86. DOI: 10.1016/j.surfcoat.2008.08.005

[37] Romero-Gavilán F, Barros-Silva S, García-Cañadas J, Palla B, Izquierdo R, Gurruchaga M, et al. Control of the degradation of silica sol-gel hybrid coatings for metal implants prepared by the triple combination of alkoxysilanes. Journal of Non-Crystalline Solids. 2016;**453**:66-73. DOI: 10.1016/j.jnoncrysol.2016.09.026

[38] Romero-Gavilán F, Carlos-Almeida J, Cerqueira A, Gurruchaga M, Goñi I, Miranda-Salvado IM, et al. Sol-gel coatings made using methyl-modified alkoxysilanes: The balance between protection and bioactivation. Progress in Organic Coatings. 2020;**147**:105770. DOI: 10.1016/j.porgcoat.2020.105770

[39] Reyes Y, Durán A, Castro Y. Glass-like cerium sol-gel coatings on AZ31B magnesium alloy for controlling the biodegradation of temporary implants. Surface and Coatings Technology.

2016;**307**:574-582. DOI: 10.1016/j.surfcoat.2016.09.056

[40] Omar SA, Ballarre J, Castro Y, Martinez Campos E, Schreiner W, Durán A, et al. 58S and 68S sol-gel glass-like bioactive coatings for enhancing the implant performance of AZ91D magnesium alloy. Surface and Coatings Technology. 2020;**400**:126224. DOI: 10.1016/j.surfcoat.2020.126224

[41] Farag MM, Ahmed HY, Al-Rashidy ZM. Improving the corrosion resistance of magnesium alloy by magnesium phosphate/glass composite coatings using sol–gel method. Silicon. 2023;**15**(9):3841-3854. DOI: 10.1007/s12633-022-02226-0

[42] Córdoba LC, Montemor MF, Coradin T. Silane/TiO_2 coating to control the corrosion rate of magnesium alloys in simulated body fluid. Corrosion Science. 2016;**104**:152-161. DOI: 10.1016/j.corsci.2015.12.006

[43] Castro Y, Durán A. Control of degradation rate of Mg alloys using silica sol–gel coatings for biodegradable implant materials. Journal of Sol-Gel Science and Technology. 2019;**90**(1):198-208. DOI: 10.1007/s10971-018-4824-6

[44] Majumder O, Singh Bankoti AK, Kaur T, Thirugnanam A, Mondal AK. The influence of silane and silane–PMMA coatings on the in vitro biodegradation behavior of AE42 magnesium alloy for cardiovascular stent applications. RSC Advances. 2016;**6**(109):107344-107354. DOI: 10.1039/C6RA23384H

[45] Gaur S, Nigam S, Khanna AS, Raman RKS. Silane-coated magnesium implants with improved in-vitro corrosion resistance and biocompatibility. Journal of Materials Science & Surface Engineering. 2016;**4**(5):415-424. DOI: 10.1016/j.surfcoat.2016.09.021

[46] Liu X, Yue Z, Romeo T, Weber J, Scheuermann T, Moulton S, et al. Biofunctionalized anti-corrosive silane coatings for magnesium alloys. Acta Biomaterialia. 2013;**9**(10):8671-8677. DOI: 1016/j.actbio.2012.12.025

[47] Catauro M, Bollino F, Papale F, Ferrara C, Mustarelli P. Silica-polyethylene glycol hybrids synthesized by sol-gel: Biocompatibility improvement of titanium implants by coating. Materials Science and Engineering: C. 2015;**55**:118-125. DOI: 10.1016/j.msec.2015.05.016

[48] Jun SH, Lee EJ, Yook SW, Kim HE, Kim HW, Koh YH. A bioactive coating of a silica xerogel/chitosan hybrid on titanium by a room temperature sol-gel process. Acta Biomaterialia. 2010;**6**(1):302-307. DOI: 10.1016/j.actbio.2009.06.024

[49] Ballarre J, Manjubala I, Schreiner WH, Orellano JC, Fratzl P, Ceré S. Improving the osteointegration and bone-implant interface by incorporation of bioactive particles in sol-gel coatings of stainless steel implants. Acta Biomaterialia. 2010;**6**:1601-1609. DOI: 10.1016/j.actbio.2009.10.015

[50] Catauro M, Bollino F, Papale F. Surface modifications of titanium implants by coating with bioactive and biocompatible poly (ε-caprolactone)/SiO2 hybrids synthesized via sol–gel. Arabian Journal of Chemistry. 2018;**11**(7):1126-1133. DOI: 10.1016/j.arabjc.2015.02.010

[51] SMieszek A, Donesz-Sikorska A, Grzesiak J, Krzak J, Marycz K. Biological effects of sol-gel derived ZrO2 and SiO2/ZrO2 coatings on stainless steel surface - In vitro model using mesenchymal stem cells. Journal of Biomaterials Applications. 2014;**29**(5):699-714. DOI: 10.1177/0885328214545095

[52] Hernández-Escolano M, Juan-Díaz MJ, Martínez-Ibáñez M, Suay J, Goñi I, Gurruchaga M. Synthesis of hybrid sol-gel materials and their biological evaluation with human mesenchymal stem cells. Journal of Materials Science. Materials in Medicine. 2013;**24**(6):1491-1499. DOI: 10.1007/s10856-013-4900-y

[53] Juan-Díaz MJ, Martínez-Ibáñez M, Lara-Sáez I, da Silva S, Izquierdo R, Gurruchaga M, et al. Development of hybrid sol–gel coatings for the improvement of metallic biomaterials performance. Progress in Organic Coatings. 2016;**96**:42-51. DOI: 10.1016/j.porgcoat.2016.01.019

[54] Araújo-Gomes N, Romero-Gavilán F, Sánchez-Pérez AM, Gurruchaga M, Azkargorta M, Elortza F, et al. Characterization of serum proteins attached to distinct sol–gel hybrid surfaces. Journal of Biomedical Materials Research Part B: Applied Biomaterials. 2018;**106**(4):1477-1485. DOI: 10.1002/jbm.b.33954

[55] Cerqueira A, Araújo-Gomes N, Zhang Y, van den Beucken JJJP, Martínez-Ramos C, Ozturan S, et al. Evaluation of the inflammatory responses to sol–gel coatings with distinct biocompatibility levels. Journal of Biomedical Materials Research Part A. 2021;**109**(9):1539-1548. DOI: 10.1002/jbm.a.37149

[56] Martínez-Ibáñez M, Juan-Díaz MJ, Lara-Saez I, Coso A, Franco J, Gurruchaga M, et al. Biological characterization of a new silicon based coating developed for dental implants. Journal of Materials Science. Materials in Medicine. 2016;**27**(80):1-9. DOI: 10.1007/s10856-016-5690-9

[57] Araújo-Gomes N, Romero-Gavilán F, García-Arnáez I, Martínez-Ramos C, Sánchez-Pérez AM, Azkargorta M, et al. Osseointegration mechanisms: A proteomic approach. Journal of Biological Inorganic Chemistry. 2018;**23**(3):459-470. DOI: 10.1007/s00775-018-1553-9

[58] Ääritalo V, Areva S, Jokinen M, Lindén M, Peltola T. Sol-gel-derived TiO_2-SiO_2 implant coatings for direct tissue attachment. Part I: Design, preparation and characterization. Journal of Materials Science. Materials in Medicine. 2007;**18**(9):1863-1873. DOI: 10.1007/s10856-007-3062-1

[59] Areva S, Ääritalo V, Tuusa S, Jokinen M, Lindén M, Peltola T. Sol-gel-derived TiO_2-SiO_2 implant coatings for direct tissue attachment. Part II: Evaluation of cell response. Journal of Materials Science. Materials in Medicine. 2007;**18**(8):1633-1642. DOI: 10.1007/s10856-007-3064-z

[60] Fernández-Hernán JP, Torres B, López AJ, Martínez-Campos E, Rams J. Sol-gel coatings doped with graphene nanoplatelets for improving the degradation rate and the cytocompatibility of AZ31 alloy for biomedical applications. Surface and Coatings Technology. 2021;**426**:127745. DOI: 10.1016/j.surfcoat.2021.127745

[61] Romero-Gavilán F, Araújo-Gomes N, García-Arnáez I, Martínez-Ramos C, Elortza F, Azkargorta M, et al. The effect of strontium incorporation into sol-gel biomaterials on their protein adsorption and cell interactions. Colloids and Surfaces B: Biointerfaces. 2019;**174**:9-16. DOI: 10.1016/j.colsurfb.2018.10.075

[62] García-Arnáez I, Cerqueira A, Romero-Gavilán F, Elortza F, Azkargorta M, Iloro I, et al. Development and characterisation of strontium-doped sol-gel coatings to optimise the initial bone regeneration processes. Materials Today Communications. 2022;**33**:104674. DOI: 10.1016/j.mtcomm.2022.104674

[63] Omar S, Repp F, Desimone PM, Weinkamer R, Wagermaier W, Ceré S, et al. Sol–gel hybrid coatings with strontium-doped 45S5 glass particles for enhancing the performance of stainless steel implants: Electrochemical, bioactive and in vivo response. Journal of Non-Crystalline Solids. 2015;**425**:1-10. DOI: 10.1016/j.jnoncrysol.2015.05.024

[64] Ballarre J, Seltzer R, Mendoza E, Orellano JC, Mai YW, García C, et al. Morphologic and nanomechanical characterization of bone tissue growth around bioactive sol-gel coatings containing wollastonite particles applied on stainless steel implants. Materials Science and Engineering: C. 2011;**31**(3):545-552. DOI: 10.1016/j.msec.2010.11.030

[65] Cerqueira A, Romero-Gavilán F, García-Arnáez I, Martinez-Ramos C, Ozturan S, Iloro I, et al. Bioactive zinc-doped sol-gel coating modulates protein adsorption patterns and in vitro cell responses. Materials Science and Engineering: C. 2021;**121**:111839. DOI: 10.1016/j.msec.2020.111839

[66] Cerqueira A, Romero-Gavilán F, García-Arnáez I, Martinez-Ramos C, Ozturan S, Izquierdo R, et al. Characterization of magnesium doped sol-gel biomaterial for bone tissue regeneration: The effect of Mg ion in protein adsorption. Materials Science and Engineering: C. 2021;**125**:112114. DOI: 10.1016/j.msec.2021.112114

[67] Romero-Gavilán F, Araújo-Gomes N, Cerqueira A, García-Arnáez I, Martínez-Ramos C, Azkargorta M, et al. Proteomic analysis of calcium-enriched sol-gel biomaterials. JBIC Journal of Biological Inorganic Chemistry. 2019;**24**(4):563-574. DOI: 10.1007/s00775-019-01662-5

[68] García-Arnáez I, Romero-Gavilán F, Cerqueira A, Elortza F, Azkargorta M, Muñoz F, et al. Correlation between biological responses in vitro and in vivo to Ca-doped sol-gel coatings assessed using proteomic analysis. Colloids Surfaces B Biointerfaces. 2022;**220**:112962. DOI: 10.1016/j.colsurfb.2022.112962

[69] Cerqueira A, García-Arnáez I, Muriach M, Azkargorta M, Elortza F, Izquierdo R, et al. The effect of calcium-magnesium mixtures in sol-gel coatings on bone tissue regeneration. Biomaterials Science. 2022;**10**(19):5634-5647. DOI: 10.1039/d2bm00742h

[70] García-Arnáez I, Romero-Gavilán F, Cerqueira A, Azkargorta M, Elortza F, Suay J, et al. Proteomics as a tool to study the osteoimmunomodulatory role of metallic ions in a sol-gel coating. Journal of Materials Chemistry B. 2023;**11**(34):8194-8205. DOI: 10.1039/d3tb01204b

[71] Martinez-Ibañez M, Aldalur I, Romero-Gavilán FJ, Suay J, Goñi I, Gurruchaga M. Design of nanostructured siloxane-gelatin coatings: Immobilization strategies and dissolution properties. Journal of Non-Crystalline Solids. 2018;**481**:368-374. DOI: 10.1016/j.jnoncrysol.2017.11.010

[72] Araújo-Gomes N, Romero-Gavilán F, Lara-Sáez I, Elortza F, Azkargorta M, Iloro I, et al. Silica–gelatin hybrid sol–gel coatings: A proteomic study with biocompatibility implications. Journal of Tissue Engineering and Regenerative Medicine. 2018;**12**(7):1769-1779. DOI: 10.1002/term.2708

[73] Cerqueira A, Romero-Gavilán F, Araújo-Gomes N, García-Arnáez I, Martinez-Ramos C, Ozturan S, et al. A possible use of melatonin in the dental field: Protein adsorption and in vitro cell response on coated titanium. Materials Science and Engineering: C. 2020;**116**:111262. DOI: 10.1016/j.msec.2020.111262

[74] Radin S, Ducheyne P. Controlled release of vancomycin from thin sol-gel films on titanium alloy fracture plate material. Biomaterials. 2007;**28**(9):1721-1729. DOI: 10.1016/j.biomaterials.2006.11.035

[75] Toirac B, Garcia-Casas A, Monclús MA, Aguilera-Correa JJ, Esteban J, Jiménez-Morales A. Influence of addition of antibiotics on chemical and surface properties of sol-gel coatings. Materials. 2022;**15**:4752. DOI: 10.3390/ma15144752

[76] Toirac B, Aguilera-Correa JJ, Mediero A, Esteban J, Jiménez-Morales A. The antimicrobial activity of micron-thin sol–gel films loaded with linezolid and cefoxitin for local prevention of orthopedic prosthesis-related infections. Gels. 2023;**9**:176. DOI: 10.3390/gels9030176

[77] Toirac B, Garcia-Casas A, Cifuentes SC, Aguilera-Correa JJ, Esteban J, Mediero A, et al. Electrochemical characterization of coatings for local prevention of Candida infections on titanium-based biomaterials. Progress in Organic Coatings. 2020;**146**:105681. DOI: 10.1016/j.porgcoat.2020.105681

[78] Romero-Gavilán F, García-Arnáez I, Cerqueira A, Scalschi L, Vicedo B, Villagrasa A, et al. Insight into the antibacterial mechanism of Cu-enriched sol-gel coatings employing proteomics. Biomaterials Science. 2023;**11**:1042-1055. DOI: 10.1039/D2BM01421A

[79] Balestriere MA, Schuhladen K, Herrera Seitz K, Boccaccini AR, Cere SM, Ballarre J. Sol-gel coatings incorporating borosilicate bioactive glass enhance anti corrosive and surface performance of stainless steel implants. Journal of Electroanalytical Chemistry. 2020;**876**:114735. DOI: 10.1016/j.jelechem.2020.114735

[80] Molina J, Valero-Gómez A, Pocoví-Martínez S, Ibiza-Palacios MS, Bosch F. Simple and effective sol-gel methodology to obtain a bactericidal coating for prostheses. Journal of Sol-Gel Science and Technology. 2023;**108**(3):809-826. DOI: 10.1007/s10971-023-06237-0

[81] Palla-Rubio B, Araújo-Gomes N, Fernández-Gutiérrez M, Rojo L, Suay J, Gurruchaga M, et al. Synthesis and characterization of silica-chitosan hybrid materials as antibacterial coatings for titanium implants. Carbohydrate Polymers. 2019;**203**:331-341. DOI: 10.1016/j.carbpol.2018.09.064

[82] Ballarre J, Aydemir T, Liverani L, Roether JA, Goldmann WH, Boccaccini AR. Versatile bioactive and antibacterial coating system based on silica, gentamicin, and chitosan: Improving early stage performance of titanium implants. Surface and Coatings Technology. 2020;**381**:125138. DOI: 10.1016/j.surfcoat.2019.125138

[83] García-Arnáez I, Palla B, Suay J, Romero-Gavilán F, García-Fernández L, Fernández M, et al. A single coating with antibacterial properties for prevention of medical device-associated infections. European Polymer Journal. 2019;**113**:289-296. DOI: 10.1016/j.eurpolymj.2019.02.002

Chapter 3

Proteic Sol-Gel Route: An Eco-Friendly Method to Synthesize Magnetic Nanoparticles

Bárbara Costa, Sílvia R. Gavinho, Susana Devesa, Manuel Almeida Valente, Manuel P.F. Graça and Sílvia Soreto

Abstract

The sol-gel method is a well-established technique for synthesizing metal oxide nanoparticles, widely employed in both research and industrial applications. However, significant drawbacks, such as the use of potentially toxic organic solvents, the lengthy synthesis time, and the high cost of precursors, have driven the scientific community to explore and develop alternative routes to optimize and improve the Sol-Gel method. In this line, the proteic sol-gel route has emerged as a sustainable eco-friendly method that utilizes natural proteins, as both stabilizing and templating agents during the synthesis process. This alternative is already used for the synthesis of magnetic nanoparticles and is recognized in the literate as an efficient method for obtaining magnetic nanoparticles with unique properties, including controlled size distribution, improved thermal stability, and high magnetic response. This chapter provides a comprehensive review of the proteic sol-gel route and presents a case study on the synthesis of an innovative magnetic composite consisting of gadolinium ferrite ($Gd_3Fe_5O_{12}$) and zinc ferrite ($ZnFe_2O_4$) using an eco-friendly proteic method using powdered coconut water. The study explores the potential applications of these nanomaterials in nanomedicine. Accordingly, the materials were extensively characterized in terms of their structural properties (thermogravimetric analysis, differential scanning calorimetry, X-ray diffraction, and Fourier transform infrared spectroscopy), morphological features (scanning electron microscopy), and magnetic behavior (vibrating-sample magnetometry and magnetic hyperthermia analysis using specific absorption rate). The resulting magnetic composite exhibited ferromagnetic behavior and demonstrated non-cytotoxic properties, highlighting its potential for biomedical applications.

Keywords: critical micelle concentration, powder coconut water, magnetic nanoparticles, composites, proteic sol-gel route

1. Introduction

Magnetic nanoparticles (MNP) are nanostructures with, at least one of the three external dimensions ranging from 1 to 100 nm, placing them within the nanoscale [1].

Their unique magnetic properties, combined with their small size, have attracted considerable significant interest. Due to their versatility, MNPs find applications across diverse fields, from physics to biology [2–5]. As a naturally abundant component of the universe, they are found almost everywhere, from intergalactic space to living organisms. Notably, the human brain contains approximately 10^8 magnetic nanoparticles per gram of tissue [6, 7]. Magnetic nanoparticles can be classified into five categories based on their response to an applied magnetic field: diamagnetic, paramagnetic, ferromagnetic, ferrimagnetic, and antiferromagnetic [8].

Diamagnetic MNP lack permanent magnetic dipoles in the absence of an external field, but when exposed to a magnetic field, they develop dipoles-oriented opposite direction to the applied field.

In contrast, paramagnetic nanoparticles possess intrinsic magnetic dipoles, but these dipoles remain randomly oriented without an external magnetic field. When subjected to a magnetic field, the dipoles align in its direction.

Ferromagnetic nanoparticles exhibit permanent net magnetic dipole moments even in the absence of an external magnetic field. In contrast, ferrimagnetic and antiferromagnetic nanoparticles present antiparallel dipole arrangements. Ferrimagnetic nanoparticles retain a net magnetic moment due to unequal opposing dipoles, whereas antiferromagnetic nanoparticles have equal and opposite dipoles resulting in zero overall magnetization [8].

In biomedical applications, the most relevant types of MNP are ferromagnetic, ferrimagnetic, and superparamagnetic nanoparticles. Superparamagnetism is a phenomenon observed in nanoscale ferromagnetic and ferrimagnetic particles, where thermal fluctuations prevent permanent magnetization in the absence of an external field [7]. These nanoparticles can rapidly switch their magnetic state in response to an applied external magnetic field, and they exhibit magnetization only when an external field is applied, without retaining residual magnetization or coercivity [5, 8]. Due to their strong and reversible magnetic response, superparamagnetic nanoparticles are widely used in various applications, including medical imaging techniques (e.g., magnetic resonance imaging (MRI)), as well as in drug delivery and hyperthermia treatments [2, 8]. Lastly, the blocking temperature (TB) is the threshold above which ferromagnetic and ferrimagnetic nanoparticles transition to superparamagnetic behavior. In addition to magnetic behavior, as saturation magnetization (Ms), remanent magnetization (Mr), coercivity (Hc), and blocking temperature (TB), several other parameters are crucial for characterizing MNP, including size, shape, composition, and crystal structure.

Saturation magnetization (Ms) represents the maximum magnetization an MNP can achieve under an external magnetic field. Remanent magnetization (Mr) refers to the magnetization that remains after the external magnetic field is removed. Coercivity (Hc) is the intensity of the external magnetic field required to reduce the magnetization to zero and is also known as the width of the M-H hysteresis loop; Hc is highly dependent on particle size [7].

Since their discovery, the exceptional magnetic properties of MNP have been largely attributed to their high surface-to-volume ratio [8]. While various correlations exist between nanoparticle size and magnetic properties, the most notable is the transition between single-domain and superparamagnetic behavior [5].

At the macroscopic level, magnetic materials consist of microscopic crystalline grains, each acting as a single crystal with its own preferred magnetization axis. To minimize magnetostatic energy, these materials typically form a multi-domain state, where they spontaneously divide into multiple magnetic domains separated by domain walls. Each domain has uniform magnetization in a specific direction, but the orientations of different domains vary randomly [9].

Domain walls are considered defects in the material due to their significant width, which requires energy to form. The formation of magnetic domains is governed by the balance between domain wall energy (which increases proportionally with the interface area of domain walls) and magnetostatic energy (which scales proportionally with material volume) determining domain formation [5]. As particle size decreases, the system favors a uniform magnetization state with the lowest energy, resulting in a single-domain configuration [5]. The system stabilizes at a critical size, D_{crit} [9].

The critical single-domain size (D_{crit}) depends on several factors, including exchange stiffness, saturation magnetization (Ms), crystal anisotropy, particle shape, and domain-wall energy [5, 9, 10]. Although D_{crit} can vary by material, it typically falls within the range of 10–100 nm [9]. The second important size-related constraint to consider is the superparamagnetic limit. Superparamagnetism occurs in small nanoparticles that exhibit ferrimagnetic or ferromagnetic behavior. Because these nanoparticles are single-domain particles, the stability of their magnetic moment in a specific direction is strongly dependent by magnetic anisotropic energy per particle [5].

MNP can be prepared through chemical, physical, and biological approaches, each offering distinct advantages for tailoring nanoparticles properties to specific applications. Among chemical synthesis techniques, the sol-gel method stands out for its ability to produce large volumes of high-purity MNP with the excellent quality. Unlike traditional methods, the sol-gel is a bottom-up approach that enables the synthesis of nanocomposites at relatively lower temperatures, making it a more efficient and versatile option.

The sol-gel process begins with the formation of colloidal dispersion, the sol, which undergoes chemical transformations to form a gel—a three-dimensional network created through irreversible chemical reactions [11]. This method primarily involves two key reactions: hydrolysis and polycondensation. Initially, metal undergo hydrolysis generating reactive intermediates that subsequently undergo polycondensation, leading to sol formation. As the process continues, the sol evolves into a wet gel containing solvents. The final gel product is obtained through a drying process, during which solvent is evaporate leaving behind a porous solid material [11, 12]. To further enhance its structural and functional properties, the gel is subjected to heat treatment processes (calcination and/or sintering), optimizing its composition and physical stability.

One of the key advantages of the sol-gel method is its precise control over the size, composition, and shape of MNP. However, this process is often time-consuming and may require the use of toxic organic solvents [13]. As a result, the scientific community has been actively investigating alternative approaches that preserve the benefits of the sol-gel method while mitigating its environmental and health hazards, processing times, and costs.

2. Proteic sol-gel route

The proteic sol-gel route follows the same fundamental steps and processes of the sol-gel method, differing primarily in the choice of polymeric agent. Instead of conventional precursors like metal alkoxides, this method utilizes natural alternatives such as coconut water [14–16], gelatin [17–19], whey protein [20], and agar-agar [21]. Compared to the sol-gel method, the proteic route has several advantages: employs safer solvents; ensures precise control over particles size; and provides excellent chemical homogeneity. Additionally, its reliance on natural, non-toxic materials makes it a more environmentally friendly and sustainable synthesis method [17, 22–24].

Gelatin is widely studied in the literature as a polymerizing agent, either used alone or in combination with others. Azevêdo et al. [17] successfully synthesize cobalt tungsten nanoparticles for luminescent sensors. Ferreira et al. [18] employed commercial flavorless gelatin to synthesize spinel ferrite nanoparticles including $NiFe_2O_4$, $CoFe_2O_4$, and $CuFe_2O_4$. Ferraz et al. [19] employed gelatin as a ligand to produce an innovative ceramic pigment, β-$LiAlSi_2O_6$ [19].

Gelatin is a well-established polymerizing agent for the proteic sol-gel method. As a renewable organic ligand, it contains high concentrations of partially hydrolyzed collagen, enhancing its functionality [19, 25]. To ensure optimal performance, slow heating rates during the drying and calcination steps are essential. This controlled approach helps prevent material shrinkage during solvent removal and preserves the integrity of organic cross-linked polymeric chains [19, 20, 25]. Another promising polymeric agent that can be considered in the sol-gel synthesis is the coconut water, which is the aqueous part of the coconut endosperm [26]. Since its introduction in the 1940s, coconut water has been extensively studied. It consists of water, inorganic compounds, and organic substances [27], including sugars, minerals, proteins, free amino acids, and growth factors—key components that facilitate the polymerization reaction [28]. These compounds act as a suspension stabilizers, surfactants, and nucleation media, playing an important role in the conduction and assembly of MNP [27]. The coconut-water assisted sol-gel method has already been described as a successful method to produce different types of MNP, such as $NiFe_2O_4$ [29], $LiFe_5O_8$ [27], $Y_2O_3{:}Eu^{3+}$ [28], $Y_2O_3{:}Nd^{3+}$ [30], niobium oxides [31], $SrFe_{12}O_{19}$ [32], $BaFe_{12}O_{19}$ [33], $CoFe_2O_4$ [14], $CaFe_2O_4$ [34], and Fe_3O_4 [15].

The proteic sol-gel route stands out as eco-friendly alternative to conventional sol-gel methods. With a diverse range of natural polymerizing agents, including coconut water, gelatin and whey, this green synthesis alternative retains the advantages of the traditional sol-gel method, while eliminating its inherent toxicity.

3. Critical micelle concentration

The critical micelle concentration (CMC) is defined as the minimum concentration of surfactant required for micelle formation [32]. Surfactants are amphiphilic molecules composed of two distinct regions: a hydrophilic (water-attracting) head and a hydrophobic (water-repelling) tail [33].

When surfactant is added to a solution at increasing concentrations, the molecules initially accumulate at the interface, reducing the systems free energy by lowering the surface tension. This occurs as the hydrophobic tails are removed from direct contact with water. Once the surface becomes saturated with surfactant molecules, further addition leads to the self-assembly of surfactant molecules into micelles. This aggregation further minimizes the systems free energy by reducing the contact between water and the hydrophobic components. Beyond this saturation point, any additional surfactant introduced into the system contributes to micelle formation rather than further surface accumulation. **Figure 1** illustrates the micelle formation process [32].

Different physicochemical properties change as the CMC is reached, allowing for a variety of calculation techniques. In particular, properties such as electrical and thermal conductivity, surface tension, and viscosity change drastically when the CMC is reached.

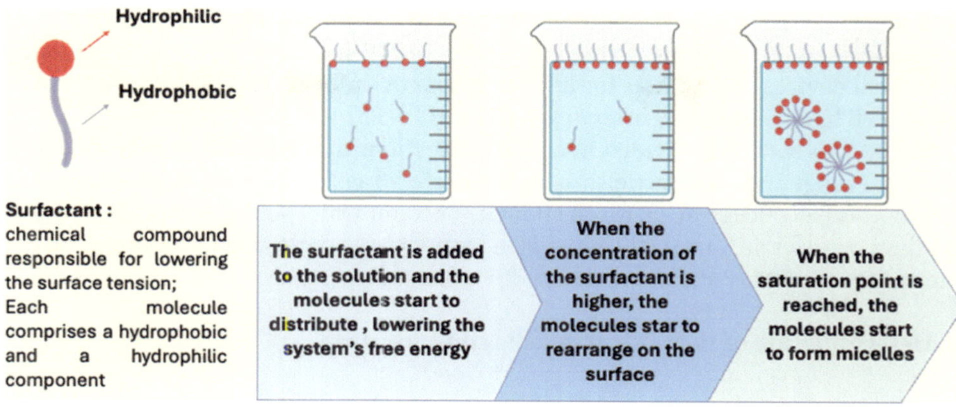

Figure 1.
Representation of the process of formation of micelles. Created in BioRender, Costa B. (2025).

Nesmerák and Nemcová [35] claim that there are two categories into which the techniques used to evaluate CMC may be divided: direct and indirect. Indirect approaches look at how a substance added to the surfactant solution changes as a function of the surfactant concentration, whereas the direct look at how a property of the surfactant solution changes in response to concentration variations. Using the direct method, the CMC is calculated directly through the discontinuity or slope variation associated with the property under study, for instance, electrical conductivity, refractive index and viscosity. Some examples of the indirect approach are the voltametric and spectrometric methods [35, 36].

In the proteic sol-gel route, a natural protein supply serves as a surfactant. Calculating the CMC is therefore crucial and required in order to comprehend the optimal circumstances under which the natural surfactant should be employed. Conductance measurements [37], UV-vis spectroscopic method [38], evaluation of the surface tension [39], and resonance Rayleigh scattering technology [40] are some of the techniques that can be used to determine the CMC concentration. Despite using different mechanisms to determine CMC, the basic concept is equitable: the analysis of the abrupt variation verified at physical properties resulting from the micelle formation [38].

4. Proteic sol-gel route with powder coconut water to synthesize magnetic nanoparticles

The ecological advantages of using a proteic sol-gel method with a natural surfactant, zinc ferrite and gadolinium iron garnet were synthesized using powder coconut water (PCW). Subsequently, after obtaining the best phase of each compound, zinc ferrite and gadolinium iron garnet were mixed to obtain a composite, suitable to be used for biomedical applications.

Therefore, thermal, structural, morphological, magnetic, and biological characterizations were performed. For the thermal analysis, differential thermal analysis (DTA), thermogravimetric analysis (TG), and X-ray diffraction (XRD) and Fourier-infrared transformed spectroscopy (FTIR) were conducted to the structural analysis. Regarding the morphological characterization, scanning electronic microscopy (SEM) was used. A vibrating sample magnetometer (VSM) provided the magnetic behavior of the samples, while the specific absorption rate (SAR) value characterizes the viability of the samples to be used for magnetic hyperthermia. Biologically, cytotoxicity assays according to the ISO 10993-5:2009 standard "Biological Evaluation of medical devices-Part5: Tests for in vitro cytotoxicity" were performed using the SaOs-2 cell line (human osteosarcoma, ATCC HTB-85) [41].

Hence, this section is divided into four subsections that guide the synthesis process of magnetic nanoparticles using a proteic sol-gel route with PCW. Firstly, the critical micelle concentration for the powder coconut water solution was determined, and then, zinc ferrite and gadolinium iron garnet were synthesized. Additionally, the composite was analyzed.

4.1 Determination of the critical micelle concentration

Different magnetic nanoparticles have been synthesized using the proteic sol-gel with coconut water [14–16]. As referred in Section 2, the coconut water acts as a surfactant. Considering that the aim is to synthesize magnetic nanoparticles using the sol-gel method, it is essential to determine the critical micellar concentration of the PCW that will be used for the synthesis process. This determination is crucial, since the synthesis process can only be verified if the concentration of the PCW is, at less, equal to the CMC value.

To determine the CMC of the PCW, electrical measurements, more specifically, the electrical conductivity of the solution, were analyzed. As illustrated in **Figure 2**, the experimental setup comprises an impedance analyzer and a dielectric measurement cell, containing two stainless steel electrodes and the PCW solution. The electrodes were immersed in the PCW solution. Different solutions of CMC were prepared: [PCW] = {0.1; 0.2; 0.3; 0.4; 0.5; 0.55; 0.6; 0.65; 0.7; 0.75; 0.8; 0.9; 1} mol. dm^{-3}. The analysis was conducted under room temperature, in a range of frequencies from 500Hz to 1 MHz.

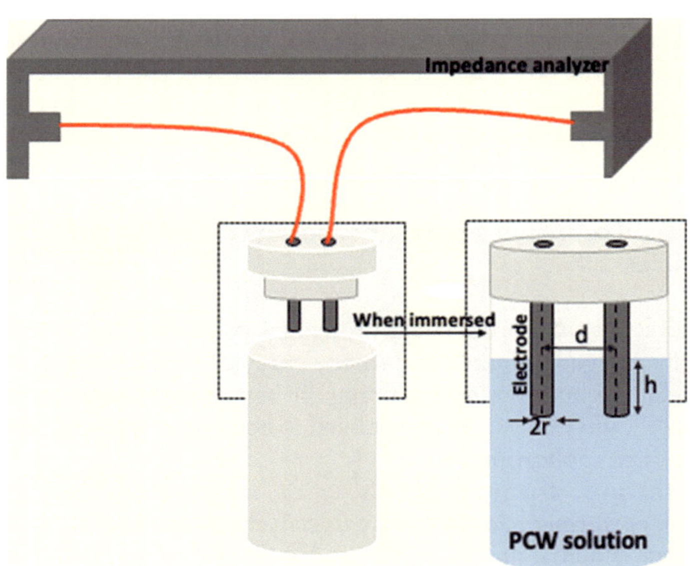

Figure 2.
Schematic representation of the used setup to determine the CMC. An impedance analyzer was connected to electrodes immersed in the PCW solution. The figure represents the electrodes with a radius of r, with a distance d between them, and h as the length of the electrodes submerged in the PCW solution.

The measures obtained with the impedance analyzer are equivalent to a resistance, R_p, in parallel with a capacitor, C_p. Therefore, it is possible to calculate the real part of the complex permittivity, often known as dielectric constant, ε', (Eq. (1)), and the imaginary part of the complex permittivity known as dielectric losses, ε" (Eq. (2)), and the solution conductivity, σ (Eq. (3)).

$$\varepsilon' = \frac{C_p}{\pi\varepsilon_0 \text{Th}_{,}} \text{Th} \left[\frac{d}{} + \sqrt{\left(\frac{d}{}\right)^2 -} \right] \tag{1}$$

$$\varepsilon'' = \frac{1}{\aleph_0 R_p \ h} \text{Th}_{,} \left[\frac{d}{r} + \sqrt{\left(\frac{d}{r}\right)^2 -} \right] \tag{2}$$

$$\sigma = 2\pi\omega\varepsilon''\varepsilon_0 \tag{3}$$

where, respectively, ε_0 corresponds to the dielectric constant of vacuum, d is the distance between the two electrodes, r is the radius of the electrodes, and h represents the length of the electrodes submerged in the PCW solution [42].

Substituting the values obtained with the impedance analyzer at Eqs. (1) and (2), C_p and R_p it was possible to obtain the values for ε' and ε''. Subsequently, the conductivity was calculated. **Figure 3** represents the variation of the conductivity as a function of the concentration of the PCW solution.

Observing **Figure 3**, the conductivity of the solution increases with the concentration of the PCW solution. As stated before, the CMC can be calculated through the changing on the slope of the property under study. Therefore, a frequency of 10^5Hz was chosen to validate this variation shown in **Figure 4**. The interception of the slopes leads to the critical micelle concentration, 0.59 mol.dm^{-3}. Therefore, this means that for the sol-gel reaction to occur, the concentration of PCW must be equal to or greater than 0.59 mol.dm^{-3}.

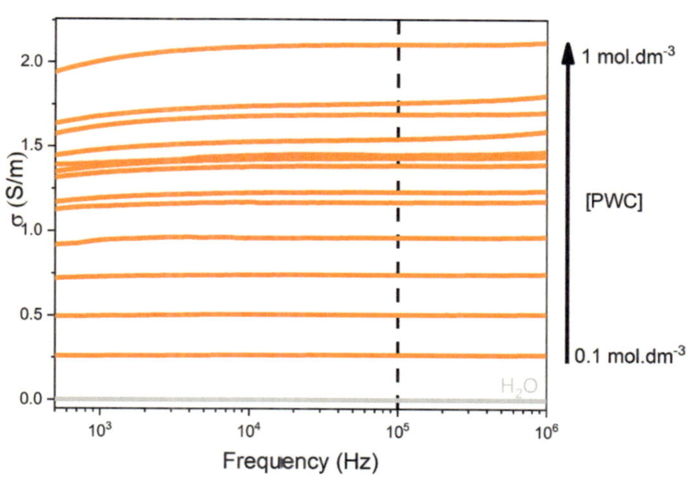

Figure 3.
Variation of the conductivity with the frequency, for the different concentrations of PCW.

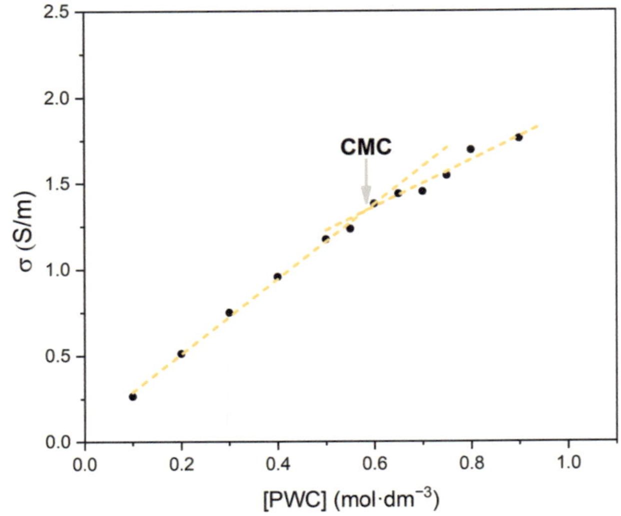

Figure 4.
Representation of the variation of conductivity as a function of the PCW concentration. With the variation of the slope of conductivity, it was possible to calculate the CMC value.

4.2 Synthesis of zinc ferrite by PCW sol-gel method

$ZnFe_2O_4$ powders were produced using as precursors iron (III) nitrate ($Fe(NO_3)_3{\cdot}9H_2O$) and zinc nitrate hexahydrate ($Zn(NO_3)_2{\cdot}6H_2O$) (Merck KGaA, Darmstadt, Germany, ≥99.0%). Taking into account the stoichiometry of the reaction (5 $Fe(NO_3)_3{\cdot}$ $9H_2O$: 1 $Zn(NO_3)_2{\cdot}$ $6H_2O$), the precursors were previously dissolved in the PCW solution (c = 0.59 mol.dm^{-3}) and therefore mixed with a magnetic stirrer in two different steps. The first step maintained the solution temperature at 80°C within 2 hours, while in the second the solution was heated to 100°C, for 2 hours. During the synthesis process, pH and temperature measurements were taken every 30-minute period. Along the synthesis process, the solution presented significant changes, not only on the pH and temperature values, but also on the color and viscosity of the solution. After obtaining a viscous brown gel, a heat treatment of 350°C, for 1 hour, was performed to remove the solvent. Then, the dried Zn and Fe powders were subjected to 3 ton to create 10-mm-diameter pellets with, approximately, a thickness of 2 mm. The obtained pellets were then submitted to a heat treatment, using two dwell times and heating rates, according to the analysis of the DTA/TG analysis.

4.2.1 Differential thermal analysis (DTA) and Thermogravimetric (TG) analysis of the Zn and Fe powders

Figure 5 illustrates the DTA/TG results. In this type of thermograms the exothermic peaks without associated mass loss are generally associated with the formation of a crystalline structure. Therefore, the synthesize Zn- and Fe-based powders present three exothermic peaks without a significant mass loss (668, 845, and 1060°C), associated with the formation of a crystalline structure and two endothermic peaks at 400 and 1223°C, respectively. The endothermic peak around 400°C was due to the volatilization of the nitrates [34] and to the decomposition of the organic, derived from

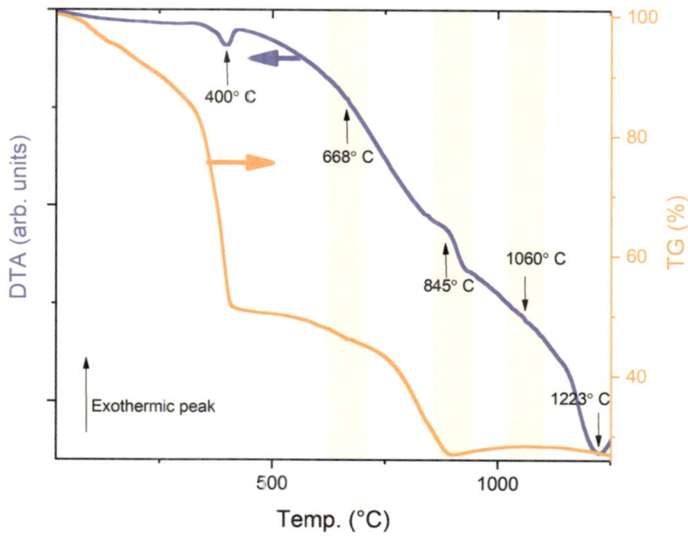

Figure 5.
DTA and TG curve of the Zn- and Fe-based powders synthesized sample using the proteic sol–gel with PCW.

the PCW [43], resulting in a mass loss of 30%, respectively, whereas the endothermic presented at 1223°C, without mass loss, can be associated with a physical transformation of the inorganic crystal phase.

Table 1 lists the different synthesized samples of Zn and Fe and their corresponding thermal treatments, enabling a simplified reading and analysis.

4.2.2 X-ray diffraction (XRD) analysis of the Zn and Fe powders

Figure 6 illustrates the diffractogram obtained for the different samples of $ZnFe_2O_4$. To facilitate a data-based comparison, **Table 2** provides a summary of the weight percentage (wt%) of each crystalline phase present in each sample, using the Rietveld refinement method. As visible in **Table 2**, the sample with highest content of $ZnFe_2O_4$ is the S5, which was heat-treated at 1200°C, for 4 h. Therefore, only S5 will be used for the subsequent analyses. According to **Figure 6**, it is visible that increasing the thermal treatment of the sample, it is possible to obtain more crystalline samples.

Considering the parameter, goodness of fit, χ^2, presents higher and closer to 1 value, indicating the good quality of the refinement [44].

Sample	Heat treatment
S1	500°C, for 4 h; heating rate of 5°C/min
S2	700°C, for 4 h; heating rate of 5°C/min
S3	850°C, for 4 h; heating rate of 5°C/min
S4	1000°C, for 4 h; heating rate of 5°C/min
S5	1200°C, for 4 h; heating rate of 5°C/min

Table 1.
Sample description with the respective heat treatment.

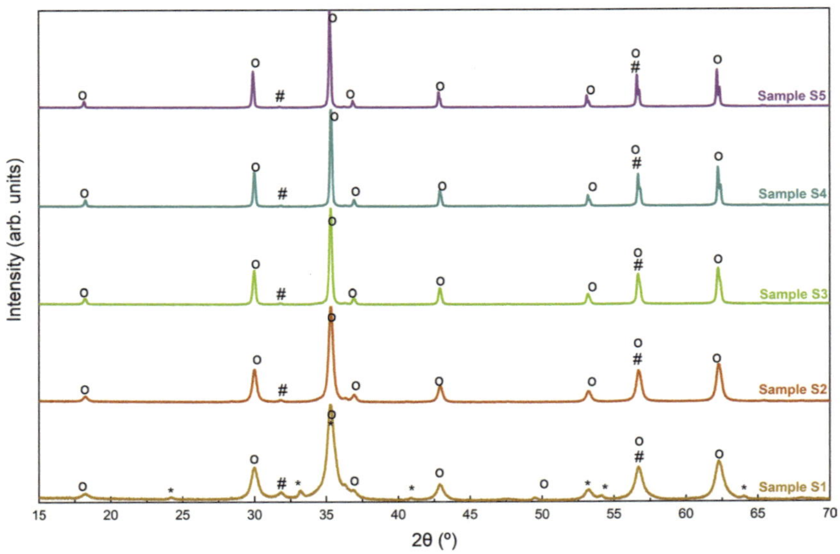

Figure 6.
*XRD Diffractograms of the samples heat-treated at 500°C (S1), 700°C (S2), 850°C (S3), 1000°C (S4) and 1200°C (S5), for 4 hours. (o: $ZnFe_2O_4$, *: Fe_2O_3; #: ZnO).*

Samples		S1	S2	S3	S4	S5
Crystalline phase	$ZnFe_2O_4$	84.9%	98.1%	98.8%	99.3%	99.6%
	ZnO	8%	1.9%	1.2%	0.7%	0.4%
	Fe_2O_3	7.2%	—	—	—	—
Goodness of fit (χ^2)		1.74	1.82	2.07	2.14	2.01

Table 2.
Quantification of the crystalline phases in each sample, based on the Rietveld refinement method outlined in related research.

4.2.3 Scanning electronic microscopy (SEM) analysis of the zinc ferrite

The morphological analysis of the sample S5 was performed (**Figure 7**). It was verified that the sample possessed agglomerates (3.8 ± 1.3 μm) with irregular shapes, such as polyhedral shapes. With ImageJ software it was possible to obtain an average particle size of 979 nm.

4.2.4 Analysis of the magnetic behavior of the zinc ferrite

Magnetically, two analyses were performed to the sample S5. Firstly, with the VSM, it was possible to characterize their magnetic behavior. Lastly, with SAR, the viability of the use of these MNPs for magnetic hyperthermia was assessed.

The hysteresis curve obtained with the VSM for the sample S5 is presented at **Figure 8**. The saturation magnetization is higher at 5 K than 300 K. With a value of M_s of 37.16 emu/g (T = 5 K) and 23.67 emu/g (T = 300 K), the sample S5 presents a ferromagnetic behavior for both situations, being in accordance with the literature [45].

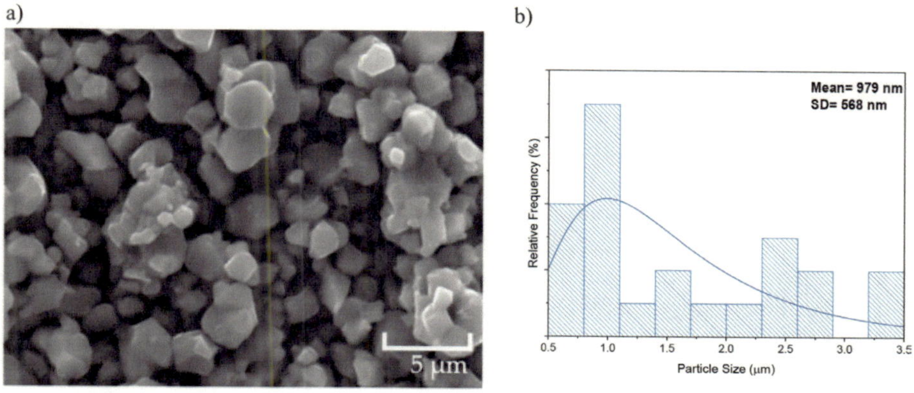

Figure 7.
(a) SEM analysis of Sample S5, and the respective (b) histogram regarding the particle size analysis.

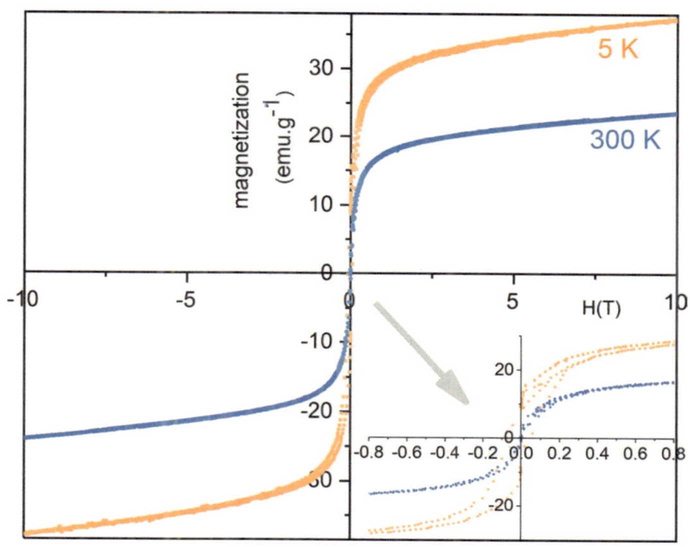

Figure 8.
Hysteresis loop obtained for sample S5, $ZnFe_2O_4$.

Regarding SAR analysis, a value of approximately 0 W/g was obtained. These results can be explained by the visible agglomeration of the nanoparticles seen in SEM analysis (**Figure 7**), resulting in a very poor thermal efficiency. Therefore, their limited thermal efficiency is significant to conclude that their applicability for magnetic hyperthermia is limited and non-advised.

4.3 Synthesis of gadolinium iron garnet with powder coconut water sol-gel method

$Gd_3Fe_5O_4$ powders were produced using as precursors iron (III) nitrate ($Fe(NO_3)_3 \cdot 9H_2O$) and gadolinium nitrate hydrate ($Gd(NO_3)_3 \cdot H_2O$) (Merck KGaA, Darmstadt, Germany, ≥99.0%). Taking into account the stoichiometry of the reaction (5 $Fe(NO_3)_3 \cdot 9H_2O$: 8 $Gd(NO_3)_2 \cdot 6H_2O$), the precursors were previously dissolved in the PCW solution (c = 0.59 mol.dm^{-3}) and therefore mixed with a magnetic stirrer in two different steps. The first step maintained the solution temperature at 80°C within 2 hours, while in the second the solution was heated to 100°C, for 2 hours. During the synthesis process, pH and temperature measurements were taken every 30-minute period. Along the synthesis process, the solution presented significant changes, not only on the pH and temperature values, but also on the color and viscosity of the solution. After obtaining a viscous brown gel, a heat treatment of 350°C, for 1 hour, was performed to remove the solvent.

Then, the dried powders were subjected to 3 ton to create 10-mm-diameter pellets with a thickness of 2 mm, approximately. The obtained pellets were then submitted to an heat treatment, using two dwell times and heating rates, according to the analysis of the DTA/TG analysis.

4.3.1 Differential thermal analysis (DTA) and thermogravimetric (TG) analysis of the gadolinium iron garnet

The DTA and TG analyses of the powders were performed under nitrogen atmosphere with a 200 mL flux, with a heating rate of 5°C/min, from room temperature to 1400°C.

Figure 9 illustrates the obtained results. In this type of thermograms, the exothermic peaks with no mass loss associated are generally linked to the formation of a crystalline structure. Therefore, the synthesize Gd- and Fe-based powders present two exothermic bands, without a significant mass loss, centered at 1223 and 1400°C, respectively. As these bands can be associated with the formation of a crystalline phase, so they will be chosen as temperatures for the thermal treatment. The exothermic band centered at 705°C with an associated mass loss is correlated with interactions of the sample with the atmosphere that results in combustion processes.

Table 3 resumes the samples and their condition of thermal treatment. To study the influence of the time of heat-treatment on the crystallinity of the Gd and Fe powders, two different times were used for the heat-treatment performed at 1200°C, 4 h and 24 h.

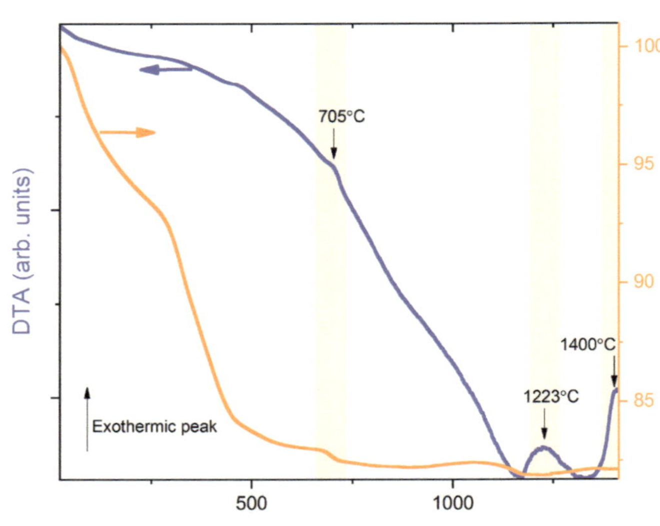

Figure 9.
Figure 5 *DTA and TG curve of the Gd- and Fe-based powders synthetized using the proteic sol–gel with PCW.*

Sample	Heat treatment
S6	1200°C, for 4 h; heating rate of 5°C/min
S7	1200°C, for 24 h; heating rate of 5°C/min
S8	1400°C, for 4 h; heating rate of 5°C/min

Table 3.
Sample description with the respective heat treatment.

4.3.2 X-ray diffraction (XRD) analysis of the gadolinium iron garnet

Figure 10 illustrates the diffractogram obtained for the samples S6, S7, and S8 of $Gd_3Fe_5O_{12}$. It is visible that none of the samples present a pure phase of $Gd_3Fe_5O_{12}$, presenting in its composition Fe_3O_4 and Fe_2O_3. To facilitate a data-based comparison, **Table 4** provides a summary of the percentage of each crystalline phase composition.

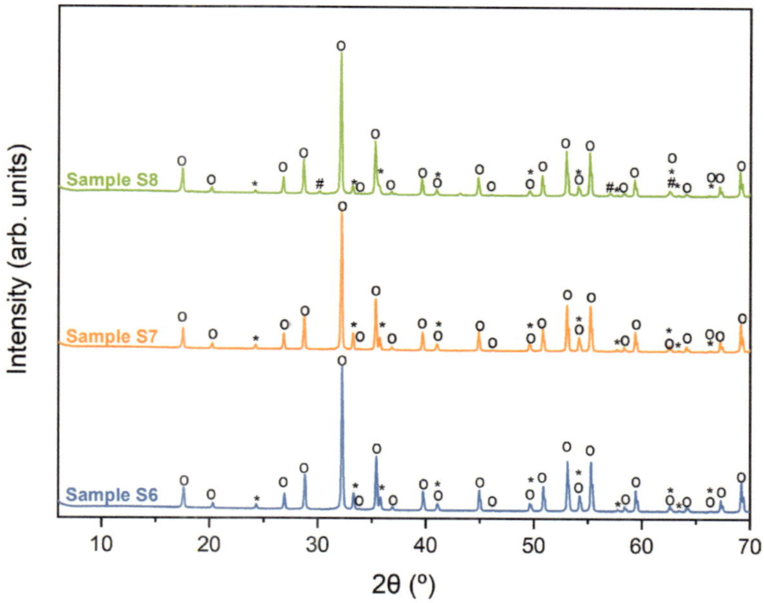

Figure 10.
*XRD diffractograms of the $Gd_3Fe_5O_{12}$ samples heat-treated at 1200°C for 4 hours (S6), 1200°C for 24 hours (S7), and 1400°C for 4 hours (S8). (o: $Gd_3Fe_5O_{12}$, *: Fe_2O_3; #: Fe_3O_4).*

Samples		S6	S7	S8
Crystalline phase	$Gd_3Fe_5O_{12}$	87%	81%	79%
	Fe_3O_4	13%	19%	5%
	Fe_2O_3	—	—	16%
Goodness of fit (χ^2)		1.78	2.15	1.98

Table 4.
Quantification of the crystalline phases in each sample, obtained from the Rietveld refinement method present in related work [16].

According to Rietveld refinement (**Table 4**), for the samples heat treated at the same temperature, S6 and S7, it was verified that a longer dwell time will promote a decrease of the $Gd_3Fe_5O_{12}$ crystal phase.

As visible in **Table 4** sample S6 is the one that presents a higher composition of $Gd_3Fe_5O_{12}$, being, therefore, the chosen one for the subsequent analysis techniques.

Considering the χ^2 parameter, that the goodness of fit, exhibits a value greater and close to 1, indicating a high-quality refinement [44].

4.3.3 Scanning electronic microscopy (SEM) analysis of the gadolinium iron garnet nanoparticles

The morphological analysis, performed by SEM, to sample S6 presented grains, with an average particle size of 1094 nm, in forms of aggregates. The structure and shape of the synthesized $Gd_3Fe_5O_{12}$ samples agreed with the described ones by Jiang et al. [46], validating the synthesis process (**Figure 11**).

4.3.4 Analysis of the magnetic behavior of the gadolinium iron garnet

In terms of magnetic characterization, the hysteresis M–H curve for the $Gd_3Fe_5O_{12}$ S6 sample at room temperature (300 K) and 5 K is shown in **Figure 12**. It is clear that the sample evaluated at 5 K shows a saturation magnetization of 57.03 emu/g, indicating a dominating ferromagnetic activity. This conduct aligns with the findings documented in the literature [47]. On the other hand, at 300 K, the sample does not saturate once the magnetization of the material increases with increasing magnetic field and does not show a magnetic hysteresis. This characteristic could be attributed to a paramagnetic behavior [46]. Analyzing the inverse of the susceptibility with temperature variation, top inset of **Figure 12**, above approximately 25 K, the sample magnetic behavior changes from ferromagnetic to paramagnetic, validating the previous analysis. The antiferromagnetic contribution from the octahedral structure of the Fe sublattice and the paramagnetic contribution linked to Gd^{3+} directly cause this temperature-dependent variation in the magnetic behavior [45]. The antiferromagnetic contribution was significant at low temperatures due to the ordered spins of the magnetic unit cell in the spin-canted FeO_6 octahedra structure. As the ambient temperature increased and the paramagnetic contribution of Gd^{3+} ions took over, the ferromagnetism progressively decreased [46].

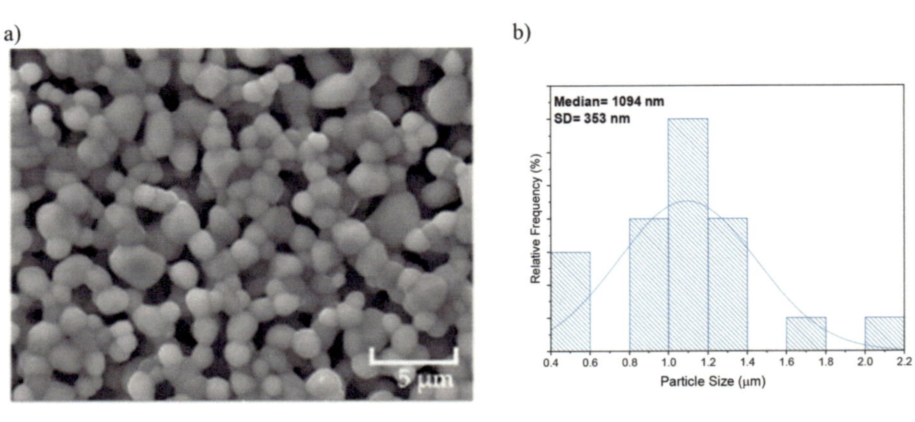

Figure 11.
a) SEM analysis of sample S6, and the respective b) histogram regarding the particle size analysis.

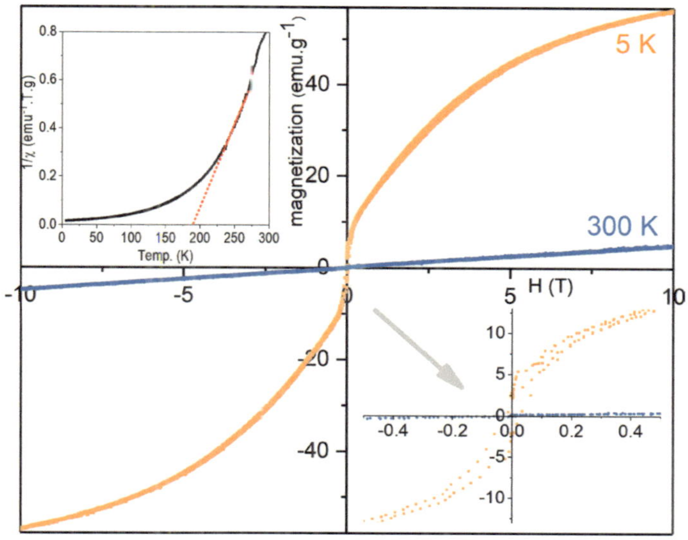

Figure 12.
Hysteresis loop obtained for Sample S6, $Gd_3Fe_5O_{12}$. Top inset in the figure is presented the inverse of the susceptibility in terms of temperature.

In terms of the SAR analysis of Sample S6, it was obtained a value of 0.3 ± 0.2 W/g. By itself, the SAR value obtained is lower than predicted, indicating a low thermal efficiency. The size of the particles, their state of aggregation, and their composition are some of the variables that might be at the root of this considerable variation.

4.4 Composite of gadolinium Iron garnet and zinc ferrite for biological applications

After obtaining the $ZnFe_2O_4$ and $Gd_3Fe_5O_{12}$ samples with the powder coconut water sol-gel method, a composite constituted with the promisor sample of the synthesized zinc ferrite, S5, and gadolinium ferrite, S6, was obtained. With the proteic sol-gel method, the achieved samples presented an average size particle higher than 100 nm, hindering their possibility to be used for biomedical applications [48].

Therefore, the first step for the synthesis of the composite was the planetary ball milling technique to reduce the size of samples S5 and S6. Following the execution of this size reduction procedure, the samples underwent a second structural and morphological analysis to assess whether the technique affected the previously obtained samples in any manner. From this second analysis, the XRD diffractograms of the $ZnFe_2O_4$ sample showed SiO_2 contamination. This contamination was only verified on the $ZnFe_2O_4$ sample, since the intensity of the peaks of the $Gd_3Fe_5O_{12}$ sample is higher than the ones from the $ZnFe_2O_4$ sample. This contamination is explained by the agate balls used for the ball milling process [49].

Thus, after reducing the size of the samples S5 and S6, the planetary ball milling was used again to mix the powders and obtain a composite (75 wt% of S6 & 25 wt% of S5).

4.4.1 Structural analysis of the composite

XRD was performed to evaluate the composition of the composite. Therefore, **Figure 13** presents the obtained diffractogram. Four different crystalline phases

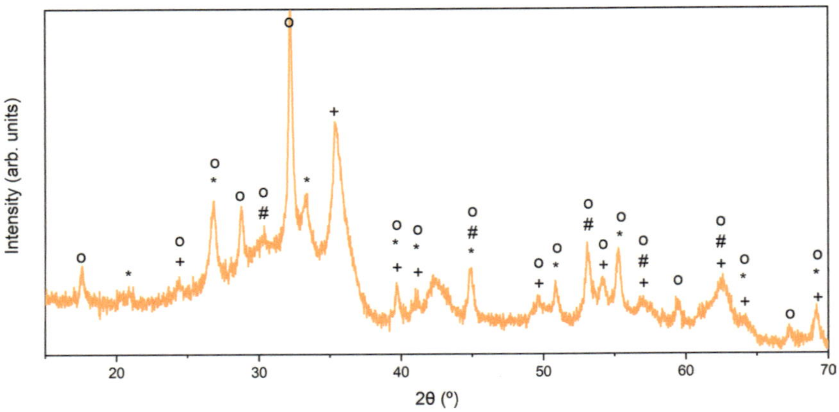

Figure 13.
*XRD diffractogram obtained for the composite (o: $Gd_3Fe_5O_{12}$, *: SiO_2, +: Fe_2O_3, #: $ZnFe_2O_4$).*

are present: $Gd_3Fe_5O_{12}$, $ZnFe_2O_4$, Fe_2O_3, and SiO_2, as expected according to previous results. At percentages levels, the composite presents a composition of 65% of $Gd_3Fe_5O_{12}$, and 10% of $ZnFe_2O_4$.

Figure 14 comprises the FTIR spectrums of S5, S6 samples and the composite. The spectrum can be divided into two well-defined regions for all the samples: Region i (400 to 1400 cm^{-1}) and Region ii (2165 to 2359 cm^{-1}). The region at highest energy is due to the metal oxides lattice vibrations, while the other one is connected with the organic functional groups bond vibrations [50]. All the samples present a vibration mode between 2165 and 2359 cm^{-1} that can be attributed to the symmetric and asymmetric starching vibration [43]. The vibration mode of the $ZnFe_2O_4$ at 1411 cm^{-1} is due to the C-C stretching vibrations [43], while 1638 cm^{-1} is related to C—O stretching

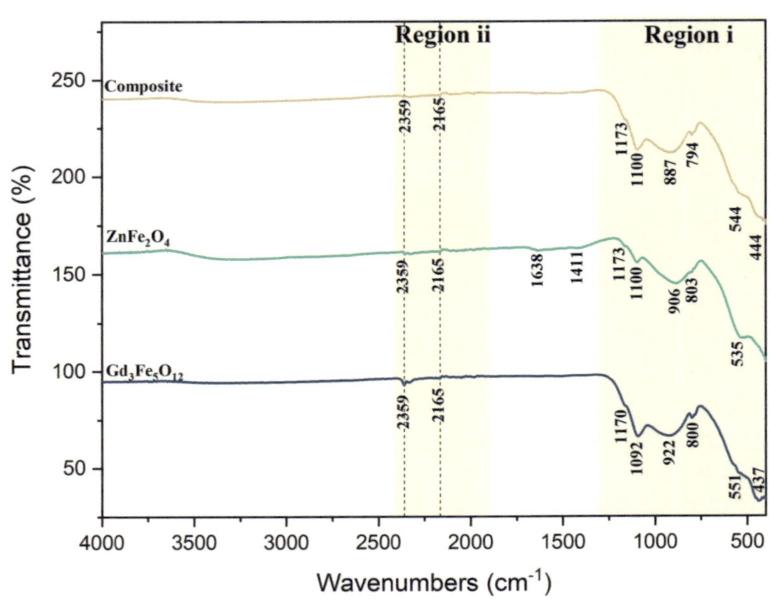

Figure 14.
FTIR Spectrum obtained for the synthesized samples S5 ($ZnFe_2O_4$), S6 ($Gd_3Fe_5O_{12}$) and the composite.

mode [50]. Regarding Region i, the vibration modes between 437 and 544 cm^{-1} are related to asymmetric stretching of the tetrahedrally coordinated Fe-O bond [51]. The vibration modes of around 1100 and 800 cm^{-1} are correlated with the Si−O−Si bond [52], corroborating the presence of SiO_2 at the synthesized samples.

4.4.2 Morphological analysis of the composite

SEM analysis was performed to samples S5 and S6 after the ball milling process and to the composite. **Figure 15** presents the obtained micrograph for a) Sample S5 and b) Sample S6 after the ball milling process. The obtained results agree with the SEM images obtained before the ball milling process, since both samples still present large agglomerates. However, in terms of average size particle, a significant decrease was verified. Before the ball milling process, Sample S5 presented an average particle size of 979 nm. After the size reduction process, the size of the $ZnFe_2O_4$ particles decreased to 196 nm. Regarding Sample S6, from the original 1094 nm it was possible to obtain $Gd_3Fe_5O_{12}$ particles with 272 nm. Therefore, the ball milling process was well-succeeded, providing a significant size reduction.

The micrography represented in **Figure 16** illustrates the SEM analysis of the composite and the analysis of the particle size. The particles of the composite present an average particle size of 155 nm, resulting from the ball-milling performed with the mixture of S5 and S6. In terms of morphology the agglomerates are still visible in a large number.

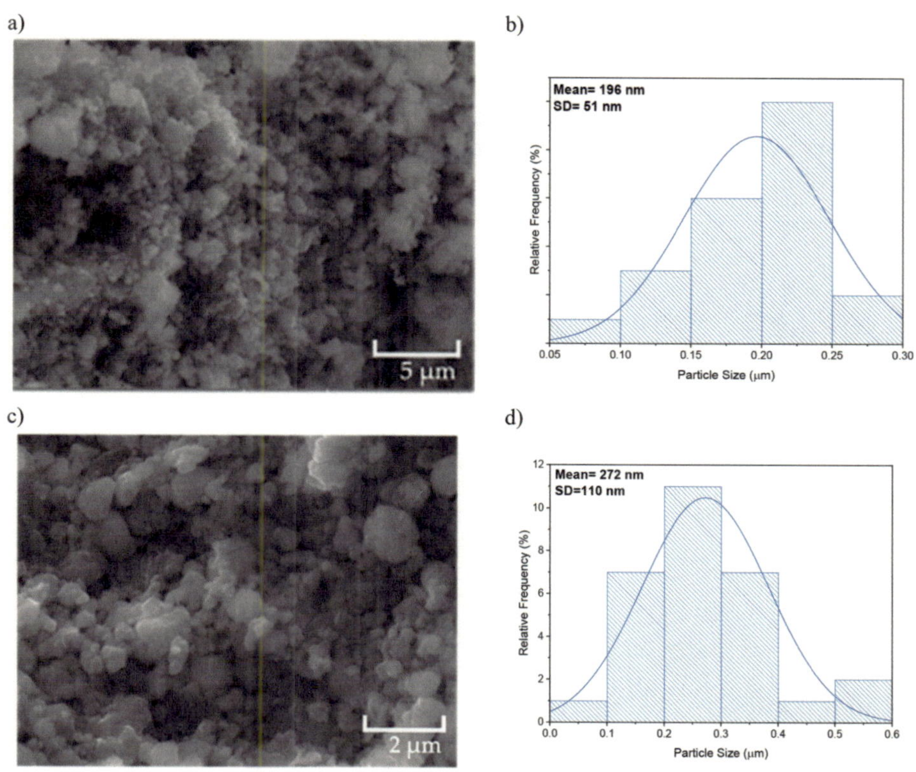

Figure 15.
SEM images obtained for a) Sample S5 and c) Sample S6 after the ball milling technique for size reduction. The histograms b) and D9 represent the analysis of the particle size for Sample S5 and S6, respectively.

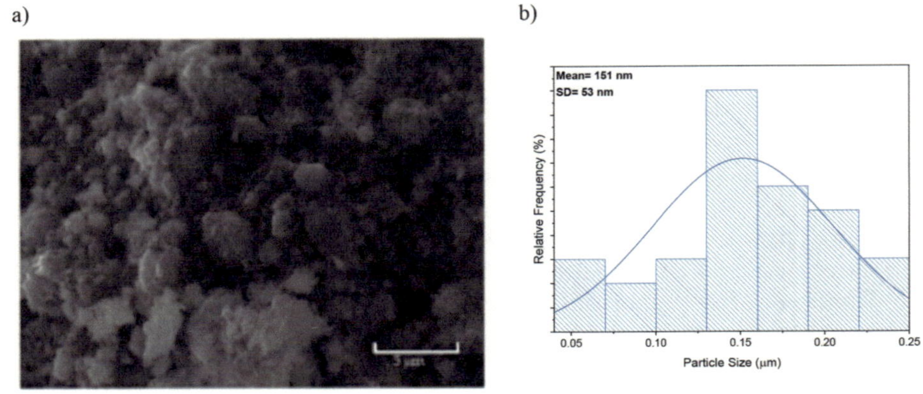

Figure 16.
SEM micrograph obtained for the synthesized composite of a) $Gd_3Fe_5O_{12}$ and $ZnFe_2O_4$ and b) the particle size analysis obtained with the illustrated analysis.

4.4.3 Magnetic characterization of the composite

Figure 17 illustrates the hysteresis loop of the composite. In agreement with the previously obtained curve M-H for the $ZnFe_2O_4$ and $Gd_3Fe_5O_{12}$ samples, the obtained saturation magnetization for the composite presents higher values at 5 K (57.30 emu/g) than at room temperature, 300 K (11.56 emu/g). Due to the presence of $ZnFe_2O_4$, the composite denotes a ferromagnetic behavior for both temperatures. This ferromagnetic behavior can be justified with the composition of the sample. The paramagnetic behavior of $Gd_3Fe_5O_{12}$ combined with the dominant ferromagnetism behavior of $ZnFe_2O_4$ and the antiferromagnetism of Fe_2O_3 [53] at ambient temperature achieves a magnetization with higher values than those of $Gd_3Fe_5O_{12}$ but lower than those of $ZnFe_2O_4$.

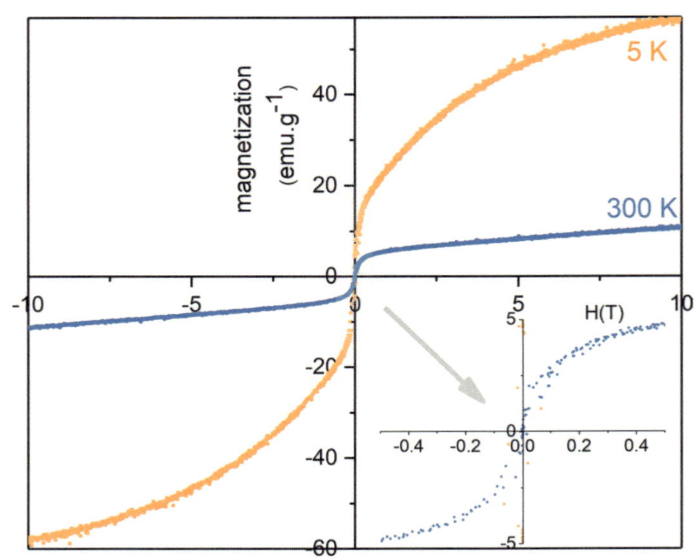

Figure 17.
Hysteresis loop obtained for the composite.

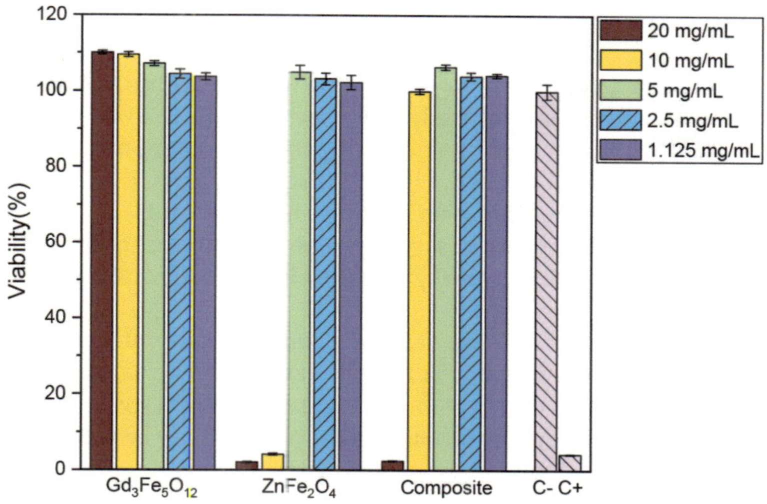

Figure 18.
Representation of the cell viability as a function of concentration of the synthesized samples. C+ and C- stands for the positive and negative control, respectively.

As stated, on the subsections 4.2.4 and 4.3.4, the SAR values obtained for the $Gd_3Fe_5O_{12}$ and $ZnFe_2O_4$ particles for itself are very low. Although the SAR obtained for the composite was 0.5 ± 0.2 (W/g), it is possible to conclude that the combination of both samples leads to an improvement of the thermal efficiency.

4.4.4 Biological analysis of the composite

One of the aims of this chapter focused on developing a composite to be used for biomedical applications through a proteic sol-gel route. Therefore, the cytotoxicity of the synthesized samples S5, S6 and the composite was evaluated (**Figure 18**). The extract method was used with different medium concentrations. Comparing all the samples, the one that presents the highest cytotoxicity is the $ZnFe_2O_4$ particles. Regarding $Gd_3Fe_5O_{12}$ particles they appear to be non-cytotoxic for all the concentrations under study. The composite, due to the presence and cytotoxicity of $ZnFe_2O_4$, presents cytotoxicity for concentrations of 20 mg/mL.

5. Conclusions

The sol-gel method is a chemical process that starts with a sol and ends with a gel, a three-dimensional structure. Although being widely used by the scientific community, it still presents serious limitations and drawbacks. From the use of toxic solvents to being a highly cost-procedure, depending on the selected precursor, the scientific community is making huge efforts to develop different strategies to overcome these drawbacks, based on green chemistry.

The proteic sol-gel route uses the bases of the sol-gel method but with natural polymers and materials as surfactants to promote the gelation process. From gelatin, whey, or coconut water, different natural protein sources are being used successfully to synthesize different compounds, magnetic nanoparticles, and other materials with this method. Using natural surfactants, the proteic sol-gel route presents serious advantages, compared to the traditional one. Being and eco-friendly and efficient solution, this method has gathered the attention of the scientific community.

During this chapter, the proteic sol-gel route using different compounds as surfactants was analyzed. Then, a case of study was presented, where a magnetic composite of $Gd_3Fe_5O_{12}$ and $ZnFe_2O_4$ was synthesized using coconut water powder as a surfactant for the proteic sol-gel method. For the synthesis process, the critical micelle concentration was determined using conductance measurements, to determine the concentration of coconut water necessary to promote the process of formation of gel. $Gd_3Fe_5O_{12}$ and $ZnFe_2O_4$ particles were synthesized individually and then mixed with a planetary ball mill.

A composite with $Gd_3Fe_5O_{12}$, $ZnFe_2O_4$, Fe_2O_3, and SiO_2 was synthesized with particles with an average dimension of 155 nm. With ferromagnetic behavior, a saturation magnetization value of 11.56 emu/g at room temperature, and a specific absorption rate of 0.5 ± 0.2 (W/g), the composite denoted its possibility to be used for biomedical applications. In terms of cytotoxicity, for concentration below 10 mg/mL the composite does not appear to be cytotoxic. This study expands the knowledge of the proteic sol-gel method using PCW as a surfactant in the synthesis of $Gd_3Fe_5O_{12}$ and $ZnFe_2O_4$ and paves the way for process optimization and its extension to other materials, namely for biomedical applications, such as medical imaging techniques (MRI).

Acknowledgements

The authors acknowledge the support of FCT—Fundação para a Ciência e a Tecnologia, I.P., in the scope of the projects LA/P/0037/2020, UIDP/50025/2020 and UIDB/50025/2020 of the Associate Laboratory Institute of Nanostructures, Nanomodelling, and Nanofabrication-i3N and UIDB/00285 – Centre for Mechanical Engineering, Materials and Processes and LA/P/0112/2020.

B. Costa acknowledges FCT—the Portuguese Foundation for Science and Technology for the PhD grant (2024.02558.BD).

Conflict of interest

The authors declare no conflict of interest.

Author details

Bárbara Costa[1], Sílvia R. Gavinho[1*], Susana Devesa[2], Manuel Almeida Valente[1], Manuel P.F. Graça[1] and Sílvia Soreto[1]

1 i3N and Department of Physics, University of Aveiro, Campus Universitário de Santiago, Aveiro, Portugal

2 Department of Mechanical Engineering, University of Coimbra, CEMMPRE, ARISE, Coimbra, Portugal

*Address all correspondence to: silvia.soreto@ua.pt

References

[1] International Organization for Standardization. ISO 80004-1:2023 Nanotechnologies - Vocabulary - Part 1: Core Vocabulary. Geneva: International Organization for Standardization; 2023

[2] Bedanta S, Barman A, Kleemann W, et al. Magnetic nanoparticles: A subject for both fundamental research and applications. Journal of Nanomaterials. 2013;**2013**:22. DOI: 10.1155/2013/952540

[3] Duguet E, Vasseur S, Mornet S, et al. Magnetic nanoparticles and their applications in medicine. Nanomedicine. 2006;**1**:157-168. DOI: 10.2217/17435889.1.2.157

[4] Tartaj P, Morales MP, Gonzalez-Carreño T, et al. Biomedical applications of magnetic nanoparticles. In: Reference Module in Materials Science and Materials Engineering. Amsterdam: Elsevier; 2016. DOI: 10.1016/B978-0-12-803581-8.02251-7

[5] Yu N, Zhao L, Cheng D, et al. Radioactive organic semiconducting polymer nanoparticles for multimodal cancer theranostics. Journal of Colloid and Interface Science. 2022;**619**:219-228. DOI: 10.1016/j.jcis.2022.03.107

[6] Dzhardimalieva GI, Eremenko IL, Elena YF, et al. Magnetic Nanoparticles. Wiley-VCH Verlag; 2009

[7] Akbarzadeh A, Samiei M, Davaran S. Magnetic nanoparticles: Preparation, physical properties, and applications in biomedicine. Nanoscale Research Letters. 2012;**7**:13. DOI: 10.1186/1556-276X-7-144

[8] Kolhatkar AG, Jamison AC, Litvinov D, et al. Tuning the magnetic properties of nanoparticles. International Journal of Molecular Sciences. 2013;**14**:15977-16009. DOI: 10.3390/ijms140815977

[9] Wu K, Su D, Liu J, et al. Magnetic nanoparticles in nanomedicine: A review of recent advances. Nanotechnology. 2019;**30**:47. DOI: 10.1088/1361-6528/ab4241

[10] Rodrigues HF, Capistrano G, Bakuzis AF. In vivo magnetic nanoparticle hyperthermia: A review on preclinical studies, low-field nano-heaters, noninvasive thermometry and computer simulations for treatment planning. International Journal of Hyperthermia. 2020;**37**:76-99. DOI: 10.1080/02656736.2020.1800831

[11] Bokov D, Turki Jalil A, Chupradit S, et al. Nanomaterial by sol-gel method: Synthesis and application. Advances in Materials Science and Engineering. 2021;**2021**:21. DOI: 10.1155/2021/5102014

[12] Sakka S. Sol-gel process and applications. In: Handbook of Advanced Ceramics. Amsterdam: Elsevier Inc.; 2013. pp. 883-910. DOI: 10.1016/B978-0-12-385469-8.00048-4

[13] Ali A, Shah T, Ullah R, et al. Review on recent Progress in magnetic nanoparticles: Synthesis, characterization, and diverse applications. Frontiers in Chemistry. 2021;**9**:629054. DOI: 10.3389/fchem.2021.629054

[14] Gomes P, Costa B, Carvalho JPF, et al. Cobalt ferrite synthesized using a biogenic sol–gel method for biomedical applications. Molecules. 2023;**28**. DOI: 10.3390/molecules28237737

[15] Jesus J, Regadas J, Costa B, et al. Green sol–gel synthesis of iron oxide

nanoparticles for magnetic hyperthermia applications. Pharmaceutics. 2024;**16**. DOI: 10.3390/pharmaceutics16121578

[16] Costa B, Carvalho J, Gavinho S, et al. Preparation and characterization of zinc ferrite and gadolinium iron garnet composite for biomagnetic applications. Materials. 2024;**17**. DOI: 10.3390/ma17122949

[17] Azevêdo HVSB, Sabino da Silva EB, Fernandes RS, et al. Green chemistry synthesis of wolframites: An investigation about stoichiometric ratios, optical, structural, vibrational, and physico-chemical properties. Journal of Cleaner Production. 2023;**385**. DOI: 10.1016/j.jclepro.2022.135528

[18] Ferreira LS, Silva TR, Silva VD, et al. Spinel ferrite MFe2O4 (M = Ni, Co, or Cu) nanoparticles prepared by a proteic sol-gel route for oxygen evolution reaction. Advanced Powder Technology. 2022;**33**:103391. DOI: 10.1016/j.apt.2021.12.010

[19] Ferraz RF, Sousa JF, Costa DDS, et al. Development of a β-LiAlSi2O6:Cr-based ceramic pigment by proteic sol-gel process using gelatin: Synthesis and characterization. Materials Research. 2022;**25**. DOI: 10.1590/1980-5373-MR-2021-0315

[20] Ferraz RF, da Conceição Costa Pereira M, RAP O. Synthesis and characterization of β-spodumene by a new sol–gel route assisted by whey protein. Journal of Sol-Gel Science and Technology. 2024;**111**:718-724. DOI: 10.1007/s10971-024-06484-9

[21] Azevêdo HVSB, Raimundo RA, Ferreira LS, et al. Green synthesis of CoWO4 powders using agar-agar from red seaweed (Rhodophyta): Structure, magnetic properties and battery-like behavior. Materials Chemistry and Physics. 2020;**242**. DOI: 10.1016/j.matchemphys.2019.122544

[22] Nair PAK, Vasconcelos WL, Paine K, et al. A review on applications of sol-gel science in cement. Construction and Building Materials. 2021;**291**. DOI: 10.1016/j.conbuildmat.2021.123065

[23] Pilban Jahromi S, Huang NM, Muhamad MR, et al. Green gelatine-assisted sol-gel synthesis of ultrasmall nickel oxide nanoparticles. Ceramics International. 2013;**39**:3909-3914. DOI: 10.1016/j.ceramint.2012.10.237

[24] Wang Y, Dong M, Guo M, et al. Agar/gelatin bilayer gel matrix fabricated by simple thermo-responsive sol-gel transition method. Materials Science and Engineering C. 2017;**77**:293-299. DOI: 10.1016/j.msec.2017.03.254

[25] de Menezes AS, Remédios CMR, Sasaki JM, et al. Sintering of nanoparticles of α-Fe2O3 using gelatin. Journal of Non-Crystalline Solids. 2007;**353**:1091-1094. DOI: 10.1016/j.jnoncrysol.2006.12.022

[26] Yong JWH, Ge L, Ng YF, et al. The chemical composition and biological properties of coconut (*Cocos nucifera* L.Water). Molecules. 2009;**14**:5144-5164. DOI: 10.3390/molecules14125144

[27] Teixeira SS, Graça MPF, Lucas J, et al. Nanostructured LiFe5O8 by a biogenic method for applications from electronics to medicine. Nanomaterials. 2021;**11**:1-15. DOI: 10.3390/nano11010193

[28] Gomes MDA, Valerio G, Macedo S. Particle size control of Y2O3:Eu3+ prepared via a coconut water-assisted sol-gel method. Journal of Nanomaterials. 2011;**2011**:1-6. DOI: 10.1155/2011/469685

[29] de Paiva JAC, Graça MPF, Monteiro J, et al. Spectroscopy studies of NiFe2O4

nanosized powders obtained using coconut water. Journal of Alloys and Compounds. 2009;**485**:637-641. DOI: 10.1016/j.jallcom.2009.06.052

[30] Gomes MA, Brandão-silva AC, Avila JFM, et al. Particle size effect on structural and optical properties of Y2O3:Nd3+ nanoparticles prepared by coconut water-assisted sol-gel route. Journal of Luminescence. 2018;**200**:43-49. DOI: 10.1016/j.jlumin.2018.04.004

[31] Soreto JMFLS, Gavinho TSR, Silva PRPCC, et al. Niobium oxide prepared by sol – Gel using powder coconut water. Journal of Materials Science: Materials in Electronics. 2019;**30**:11346-11353. DOI: 10.1007/s10854-019-01482-y

[32] Brito PCA, Gomes RF, Duque JGS, et al. SrFe12O19 prepared by the proteic sol-gel process. Physica B: Condensed Matter. 2006;**384**:91-93. DOI: 10.1016/j.physb.2006.05.159

[33] Fortes SS, Duque JGS, Macêdo MA. Nanocrystals of BaFe12O19 obtained by the proteic sol-gel process. Physica B: Condensed Matter. 2006;**384**:88-90. DOI: 10.1016/j.physb.2006.05.158

[34] Gomes HC, Teixeira SS, Graça MPF. Synthesis of calcium ferrite for energy storage applications. Journal of Alloys and Compounds. 2022;**921**. DOI: 10.1016/j.jallcom.2022.166026

[35] Nesměrák K, Němcová I. Determination of critical micelle concentration by electrochemical means. Analytical Letters. 2006;**39**:1023-1040. DOI: 10.1080/00032710600620302

[36] Massarweh O, Abushaikha AS. The use of surfactants in enhanced oil recovery: A review of recent advances. Energy Reports. 2020;**6**:3150-3178. DOI: 10.1016/j.egyr.2020.11.009

[37] Fennell D, Allen M, Ninham BW, et al. Critical micelle concentrations for Alkyltrimethylammonium bromides in water from 25 to 160~. Journal of Solution Chemistry. 1984;**13**:87-101

[38] Tanhaei B, Saghatoleslami N, Chenar MP, et al. Experimental study of CMC evaluation in single and mixed surfactant systems, using the UV-vis spectroscopic method. Journal of Surfactants and Detergents. 2013;**16**:357-362. DOI: 10.1007/s11743-012-1403-7

[39] Kerwin BA. Polysorbates 20 and 80 used in the formulation of protein biotherapeutics: Structure and degradation pathways. Journal of Pharmaceutical Sciences. 2008;**97**:2924-2935. DOI: 10.1002/jps.21190

[40] Li N, Liu S, Luo H. A new method for the determination of the first and second CMC in CTAB solution by resonance Rayleigh scattering technology. Analytical Letters. 2002;**35**:1229-1238. DOI: 10.1081/AL-120005975

[41] International Organization for Standartization. ISO 10993-5: Biological Evaluation of Medical Devices- Part 5: Tests for in Vitro Cytotoxicity. Geneva: International Organization for Standartization; 2009

[42] Tagmouti S, Bouzit SE, Costa LC, et al. Impedance spectroscopy of Nanofluids based on multiwall carbon nanotubes. Spectroscopy Letters. 2015;**48**:761-766. DOI: 10.1080/00387010.2015.1034874

[43] Satheshkumar M, Anand B, Muthuvel A, et al. Enhanced photocatalytic dye degradation and antibacterial activity of biosynthesized ZnO-NPs using curry leaves extract with coconut water. Nanotechnology for. Environmental Engineering. 2020;**5**:1-11. DOI: 10.1007/s41204-020-00093-x

[44] Toby BH. R factors in Rietveld analysis: How good is good enough? Powder Diffraction. 2006;**21**:67-70. DOI: 10.1154/1.2179804

[45] Anjaneyulu T, Raghavender A, Vijaya Kumar K, et al. Effect of particle size on the structural and magnetic properties of nanocrystalline zinc ferrite. Science, Technology and Arts Research Journal. 2014;**3**:48. DOI: 10.4314/star.v3i3.8

[46] Jiang L, Yang S, Zheng M, et al. Synthesis and magnetic properties of nanocrystalline Gd3Fe5O12 and GdFeO3 powders prepared by sol–gel auto-combustion method. Materials Research Bulletin. 2018;**104**:92-96. DOI: 10.1016/j.materresbull.2018.04.010

[47] El Dek SI, Amin RM. Dose dependent modifications in structural, magnetic and optical properties of gamma-irradiated nanocrystalline Gd3Fe5O12 garnet. Radiation Physics and Chemistry. 2023;**204**:110709. DOI: 10.1016/j.radphyschem.2022.110709

[48] Martins PM, Lima AC, Ribeiro S, et al. Magnetic nanoparticles for biomedical applications: From the soul of the earth to the deep history of ourselves. ACS Applied Bio Materials. 2021;**4**:5839-5870. DOI: 10.1021/acsabm.1c00440

[49] Nakai A, Shun'ichi T. Contamination introduced during rock sample powdering: Effects from different mill materials on trace element contamination. Geochemical Journal. 2009;**43**:389-394. DOI: 10.2343/geochemj.1.0032

[50] Manohar A, Krishnamoorthi C, Naidu KCB, et al. Dielectric, magnetic hyperthermia, and photocatalytic properties of ZnFe2O4 nanoparticles synthesized by solvothermal reflux method. Applied Physics A: Materials Science & Processing. 2019;**125**:1-10. DOI: 10.1007/s00339-019-2760-0

[51] Mallmann EJJ, Sombra ASB, Goes JC, et al. Yttrium iron garnet: Properties and applications review. Solid State Phenomena. 2013;**202**:65-96. DOI: 10.4028/www.scientific.net/SSP.202.65

[52] Liang Y, Ouyang J, Wang H, et al. Synthesis and characterization of core-shell structured SiO 2 @YVO 4:Yb 3+ ,Er 3+ microspheres. Applied Surface Science. 2012;**258**:3689-3694. DOI: 10.1016/j.apsusc.2011.12.006

[53] Vasquez-Mansilla M, Zysler RD, Arciprete C, et al. Magnetic interaction evidence in α-Fe2O3 nanoparticles by magnetization and Mossbauer measurements. Journal of Magnetism and Magnetic Materials. 1999;**204**:29-35. DOI: 10.1016/S0304-8853(99)00404-7

Chapter 4

Synthesis of Fluorine-Doped Tin Oxide Powders by Sol-Gel Combustion Method

Sara Malek

Abstract

Fluorine doped tin oxide (FTO) and other transparent conductive oxides have become increasingly popular in solar cells due to their cost-effectiveness and efficiency, leading researchers to view FTO as an encouraging option for solar cell applications. There is a necessity to create a more straightforward and effective approach for generating nanopowders with a narrow size range, as current methods are complex and require substantial organic material usage. Most existing methods involve complicated processes and the use of large amounts of organic materials. Among the many synthesis methods for nanopowders, the sol-gel combustion method and laser-assisted pyrolysis are used to produce nano-sized and homogeneous powders with the smallest grain size. The sol-gel combustion method, is simpler and more efficient for producing nanoparticles with narrow size distribution. This process offers benefits such as the use of inexpensive precursors, a simple preparation technique, and the production of nano-sized powder. In this study, the sol-gel combustion synthesis of FTO and the optimization of its effective parameters were investigated. The findings showed the production of uniform FTO nano-powders with an average particle size of 20 nm and equiaxed morphology at a precursor concentration of 0.2 molar, citric acid to precursor molar ratio of 1, and pH of 0.5.

Keywords: sol-gel combustion, synthesis, solar cell, fluorine-doped tin oxide, transparent conductive oxides

1. Introduction

Transparent conductive oxides have become popular as a base for titanium oxide photo electrodes in solar cells [1]. Thin, transparent conductive materials like fluorine-doped tin oxide (FTO), tin-doped indium oxide (ITO), and antimony-doped tin oxide are extensively used in various fields, including solar cells, due to their suitable combination of electrical, optical, and chemical properties [2]. Although ITO film has been widely utilized for many years, transparent conductive oxide films are preferred due to their lower cost and higher efficiency compared to indium [3, 4], as well as their reduced electrical resistance at elevated temperatures. Researchers have high hopes for transparent conductive oxides like FTO in solar cell applications due to its impressive qualities, including high thermal and chemical stability [3], hardness, and resistance to high temperatures [5]. Also, it has more transparency for lower resistant surface compared to antimony [6]. Both doped and undoped SnO_2 have been created using various methods, including chemical precipitation [7], hydrothermal process [8], sol-gel [9–12], and sol-gel combustion [13]. Sol-gel combustion, which combines chemical sol-gel and combustion processes [14], is known for its simplicity and efficacy in producing nanoparticles with a narrow size distribution [9]. The process offers the benefits of inexpensive precursors, a straightforward preparation technique, and resulting smallest and nano-sized powder [13]. One of the applications of FTO nano-powders is to use them in the target of sputtering method to produce transparent conductive FTO. The finer the powders, the less the energy is needed to break their atoms in the sputtering method.

In sol-gel combustion synthesis, when the initial solution is heated, it transforms into a gel with high viscosity. As the temperature increases, the gel turns into a soft and porous gray material through an exothermic combustion process. This combustion is caused by the reaction of organic materials acting as a reducing agent and nitrates or chlorides acting as an oxidizing agent. The final product is obtained by calcining at the appropriate temperature. The gel formed from sol, which occurs when excess water vapor and gases are released, does not immediately turn into ash when the temperature rises and the combustion reaction occurs. A necessary condition for the exothermic combustion process is to reach the combustion temperature before the combustion front advances throughout the gel. During the process of gel formation and drying, many gases are released from inside the gel. On the other hand, due to the presence of organic compounds inside the gel, the gel forms large bubbles similar to soap during drying, and these gases cause the bubbles to grow. Additionally, the retention of the expanded exhaust gases inside these bubbles and their growth causes the opening and weakening of the organic bonds adjacent to each other. Consequently, at one point, these bonds are separated by the rupture of the bubble and establish a connection with their neighboring ions. This localized connection causes the temperature to reach the ignition point, creating a spark at one point of the gel. This spark eventually spreads the ignition front to the rest of the gel, leaving behind powdery ash. The sol-gel combustion process is based on rocket propellant technology, using an oxidizing agent and a fuel as a regeneration agent. By controlling the proper ratio of oxidizer to fuel, high heat resulting from reactions can be achieved. It is essential to choose the appropriate ratio of oxidizer to fuel to take advantage of exothermic reactions. According to the chemical theory of propulsion, for a combustion reaction to complete, the total capacity of oxidizers must equal the total of regenerators. An additional amount of fuel is usually used to ensure the completion of the reactions. Choosing an inappropriate ratio of oxidizer to fuel can lead to the creation of unwanted intermediate phases or unreacted raw materials [4, 5].

This section discusses several studies on the synthesis of SnO_2, SnO_2 doped with F, Sb, Pt, and In_2O_3 doped with Sn and Fe nanoparticles.

Fraigi and colleagues [15] discovered a new synthesis method for tin oxide nanoparticles, in which tin metal was used as a precursor instead of tin chloride. In this research, they investigated the effect of parameters of combustion temperature, fuel amount, and calcination temperature and concluded that the combustion temperature has a strong effect on the initiation of the combustion reaction, and combustion took place at a temperature of 250°C. Nanoparticles were calcined at 500, 600, and 700°C, and the size of the crystallites increased as the calcination temperature increased.

In another work done by Neri and colleagues [16], platinum-doped tin oxide nanoparticles were synthesized using the combustion sol-gel method. They used metal tin as a precursor and investigated the effect of different fuels such as citric acid, glycine, and urea on the properties of the obtained powders. They observed that the grain size of the prepared powders depends on the nature and amount of fuel used, and the nanopowders synthesized with urea have the largest grain size, the reason for this is the higher temperature of the reaction, which with the use of urea fuel, the rate of the decomposition reaction the gel is more than citric acid.

In the work done by Chi-Hwan Han et al. [4], ITO and FTO nanoparticles were synthesized by the sol-gel combustion method using acetylene black fuel. Using Scherer's formula and considering the broadening of the diffraction peaks, the size of ITO and FTO crystals was obtained as 26 and 28 nm, respectively. The electron microscope image of the calcined powder at 750°C shows that the ITO powder contains cubic particles with a very uniform size of 26–28 nm, and the FTO powder contains round particles with a relatively wider particle size distribution of 16-38 nm.

Li et al. [17] investigated the effect of citric acid on SnO_2 particles synthesized by the hydrothermal method and found that the presence of citric acid leads to the formation of particles with a larger surface area. This shows that the effect of citric acid on the particle size of the powders depends on the synthesis method.

Sikhwivhilu et al. [18] synthesized SnO_2 nanoparticles by chemical deposition method and sol-gel method with grain size ranging from 19 to 100 nm and observed that changes in citric acid concentration directly affected grain size and specific surface area. An increase in the concentration of citric acid causes grain growth and decreases the surface area of the grain, which indicates a decrease in surface energy.

Adnan et al. [19] synthesized SnO_2 nanoparticles using the sol-gel method and examined the impact of various factors, such as ammonium hydroxide concentration and addition rate. They discovered that higher ammonium hydroxide concentrations reduced the grain surface area, and the speed of adding ammonium hydroxide had a noteworthy effect on the grain size.

Wu et al. [20] prepared SnO_2 nanoparticles doped with fluorine by sol-gel and hydrothermal methods. They used tin tetrachloride and ammonium fluoride as sources of tin and fluorine. The results show that the size of FTO nanoparticles is about 10 nm, and in comparison with tin dioxide nanoparticles, clear sol is obtained with high stability. FTO nanopowders are obtained by removing the solvent. Electrical resistance was measured by the four-point probe method. The results show that the surface resistance of FTO decreases with increasing the molar ratio of ammonium fluoride to tin in the range of 0 to 2, although increasing the molar ratio of ammonium fluoride to tin from two to five leads to an increase in surface electrical resistance. The molar ratio of fluorine to tin in FTO measured by XPS is 0.18. When the ratio of ammonium fluoride to tin in solution is 2, the surface resistance of FTO powders is 110 Ω/□.

Senthilkumar et al. [21] synthesized FTO nanoparticles by sol-gel method and then calcined them at 600 and 700°C. The results show that the particle size increases with increasing calcination temperature and the resistivity decreases to 4.61×10^{-4} Ω.cm.

A study by Jianrong Zhang and colleagues [22] successfully produced ATO nanoparticles using the nitrate-citrate combustion gel method. The research looked at the ratio of fuel to metal as a parameter. Various samples were created using different fuel ratios of citric acid to metal: 1:1, 1:2, 1:3, 1:4, 1:6, and 1:8. The highest specific level achieved was 37.5 m^2/g when the ratio was 1:6.

Yu et al. [23] successfully synthesized iron-doped In_2O_3 nanoparticles using the solution combustion method. They studied how the ratio of fuel to oxidizer affected the magnetic properties of the nanoparticles. They utilized glycine as the fuel and indium nitrate as the precursor for indium oxide. During the combustion, N_2, CO_2, and H_2O gases were released. The research indicated that the ratio of glycine to metal nitrate should be 2 for complete combustion of the metal nitrate.

Oliveira et al. [24] conducted a study on synthesizing ITO nanoparticles using the microwave solvothermal method and examined the effect of pH levels on the particles. They prepared different solutions using indium chloride tetrahydrate, tin chloride pentahydrate, and tetramethylammonium hydroxide pentahydrate precursors in ethylene glycol solvent, adjusting the pH to be both acidic and alkaline. The synthesis process took place at 200°C with a heating rate of 50°C/min and a 30-minute holding time. The results indicated that the ITO particles synthesized were spherical with a size ranging from 3.4 to 8.2 nm. An increase in pH led to an increase in particle size. The light transmittance of the synthesized material ranged from 85 to 95%, making it suitable for use in devices requiring high light transmission.

In engineering experiments, researchers commonly utilize the one factor at a time (OFAT) approach, where they adjust one factor level while keeping others constant, then observe the resulting outcome. However, this method does not allow for predicting interactions between factors, nor does it guarantee optimal results. Additionally, it presents challenges in predicting the effects of multiple factors, highlighting the significant role of statistical methods in planning, conducting, and interpreting engineering experiments [25].

Designing a reliable and valid product design becomes essential when multiple variables influence the product's properties, ensuring efficient and safe results [26].

In this research, all samples were prepared by using the D-Optimal factorial test design.

2. Experimental

This research investigated the sol-gel combustion synthesis Of FTO and the optimization of its effective parameters.

2.1 Raw materials

The experiment used materials of analytical grade. $SnCl_4{\cdot}5H_2O$ and NH_4F were used as the precursors of tin and fluorine, respectively. Citric acid was utilized as the fuel, ammonium hydroxide was employed for adjusting pH, and deionized water was used as the solvent and the hydrolysis agent.

2.2 Sample preparation

By using the D-Optimal factorial test design, the parameters related to the synthesis of FTO nanopowder were optimized in the combustion sol-gel method. For this purpose, three parameters were defined in three levels. The selected parameters were the concentration of the raw material (tin tetrachloride), the molar ratio of citric acid

		Level 1	Level 2	Level 3
Selected factor	**Units**	**(−1)[1]**	**(0)[1]**	**(1)[1]**
Concentration	Molar	0.2	0.6	1
Fuel to precursor molar ratio (portion)	–	1	2	3
pH solution	–	0.5	1.5	3

[1]*The signs in the parenthesis show the coded form of the levels.*

Table 1.
Selected factors and their levels.

to the raw material, and the pH of the solution, which were changed at three levels. **Table 1** shows the parameters and their levels, and the items in parentheses are the defined codes for factors and levels, according to Design-Expert 7 software (for more information and details about the D-optimal factorial test design method, refer to the study done by the author [27] in 2018).

To investigate the effect of three parameters and the degree of influence of each of the parameters on three levels, it is necessary to synthesize many samples. However, by using the D-Optimal test design method, the number of required samples can be reduced. In this way, with a smaller number of samples, the parameters checked differently on several samples and obtained the optimal sample condition which is the sample with the smallest particle size. According to the selected parameters and levels, **Table 2** was designed using the D-Optimal method. In **Table 2**, the third column shows the order of the experiments as determined by the Design-Expert 7 software.

In this table, 19 samples are given, which were coded for the ease of using the samples, and three Latin letters are used in their coding, the letter C indicates the concentration of the raw material and its index indicates the concentration value, the letter R indicates the ratio of fuel to the raw material And, its index represents the numerical value of this ratio, the letter P represents pH and its index represents the pH value, which was considered between 0.5 and 3.

In this research, the increase in pH was also investigated, which by increasing it to the neutral range and then alkalinity, precipitation was observed in the solution.

Based on the conditions presented in **Table 2**, the process of synthesizing nano-powder FTO is as follows: Dissolve an appropriate amount of $SnCl_4 \cdot 5H_2O$ in a suitable volume of deionized water while stirring continuously and heat it until a clear solution is obtained. Then, add a certain amount of NH_4F to adjust the fluorine concentration based on the mole ratio of [F]/[Sn] = 0.1. The following procedure was carried out: first, citric acid with an appropriate molar ratio was added to a solution as a fuel. Next, NH_4OH aqueous solution was added drop by drop under constant stirring (300 rpm) until the desired pH was achieved and the solution turned into a sol. While the solution was heated to 100°C, the sol changed to a gel. The gel was then heated until it became dry. The dried gels were further heated in a furnace at 350°C for 1 hour, resulting in their auto-combustion. Finally, the powders were calcined at 600°C for 1 hour with a heating rate of 150°C/h.

2.3 Characterization

Powder X-ray diffraction (XRD; Scifert, 3003 PTS model) was conducted to analyze the synthesized powder, including its crystal structure and phase type.

Standard order	Sample coded	Run order	Concentration	Fuel to precursor molar ratio	pH	Crystallite size (nm) (based on Scherrer equation)
1	$C_{0.2}R_2P_{0.5}$	13	0.2	2	0.5	8
2	$C_1R_3P_3$	1	1	3	3	18
3	$C_{0.2}R_1P_{1.5}$	19	0.2	1	1.5	7
4	$C_1R_2P_{0.5}$	8	1	2	0.5	13
5	$C_{0.6}R_2P_{1.5}$	10	0.6	2	1.5	16
6	$C_{0.2}R_3P_{0.5}$	3	0.2	3	0.5	11
7	$C_{0.6}R_1P_3$	9	0.6	1	3	15
8	$C_1R_3P_{1.5}$	5	1	3	1.5	18
9	$C_{0.2}R_3P_{1.5}$	14	0.2	3	1.5	14
10	$C_1R_2P_{1.5}$	12	1	2	1.5	17
11	$C_{0.2}R_2P_{1.5}$	17	0.2	2	1.5	15
12	$C_1R_1P_{0.5}$	16	1	1	0.5	9
13	$C_{0.2}R_1P_{0.5}$	15	0.2	1	0.5	5
14	$C_1R_1P_3$	11	1	1	3	16
15	$C_{0.2}R_1P_3$	2	0.2	1	3	14
16	$C_{0.6}R_3P_{0.5}$	6	0.6	3	0.5	18
17	$C_{0.6}R_2P_{0.5}$	18	0.6	2	0.5	13
18	$C_{0.6}R_3P_3$	4	0.6	3	3	18
19	$C_{0.6}R_2P_3$	7	0.6	2	3	17

[1]*Extracted from research done by the author [27].*

Table 2.
Samples prepared based on the D-Optimal design of experiments[1].

The crystallite size was calculated using the Scherrer eq. (1) and the X'pert High Score Plus analytical software.

$$D_{hkl} = \frac{0.9\lambda}{\cos\theta\sqrt{\left(B_m^2 - B_s^2\right)}} \quad (1)$$

where D_{hkl} is the average crystallite size, λ the X-ray wavelength, βm the full width at half maximum of the diffraction peak that was calculated by X'pert high score software, and βs the full width at half maximum of the diffraction peak for the standard sample which is 0.15868 and θ the Braggs diffraction angle. Particle morphology and size were observed using a field emission scanning electron microscope (FESEM; TESCAN, MIRA3 LMU), and the grain size was measured using ImageJ software, which was considered as a response in experimental designs. ImageJ software (an abbreviation of image processing and analysis in Java) is a powerful program for image analysis and is used for various applications, including grain size measurement.

The surface composition was analyzed using X-ray photoelectron spectroscopy (XPS, Bester, ESCA system) with AlK_α radiation. Fluorine content in tin dioxide was analyzed using electrochemical measurement technologies with a PF-4C (204) fluoride ion electrode instrument.

3. Results and discussion

3.1 Metallurgical characterization

Figure 1 displays the X-ray diffraction (XRD) patterns of 19 samples that were calcined at 600°C. The XRD results revealed the characteristic peaks of cassiterite structure of SnO_2 crystals, and all the peaks were well indexed to the pure phase patterns (JCPDS card no. 04-077-045). No obvious peaks of impurities, such as unreacted Sn or SnO, were observed, indicating the experiment was well designed. To check the presence of additives in materials, XPS (X-ray photoelectron spectroscopy) was used, and a graph of intensity versus binding energy is created. The presence of fluorine in the optimal sample is illustrated in **Figure 4c**, and its details have been discussed.

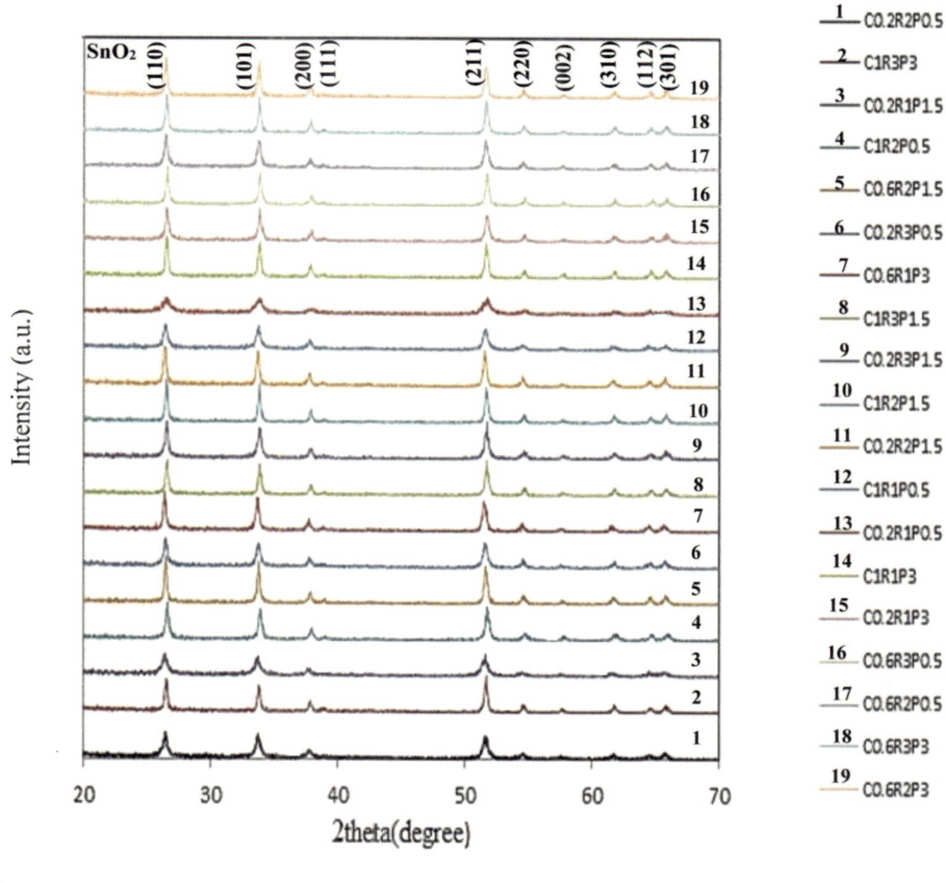

Figure 1.
XRD patterns of FTO sample (extracted from research done by the author [27]).

Using the Scherrer formula to broaden the diffraction peaks (the average of three characteristic peaks with higher intensities was considered), the average crystallite size of the tin oxide crystals is presented in **Table 2**. Due to the different conditions designed for the samples, the size of the crystallite varies, and the crystallite size of the sample with code $C_{0.2}R_1P_{0.5}$ is minimum and is 5 nm.

Figures 2 and **3** display the FESEM images of 19 samples claimed at 600°C. The particle size and morphology of the FTO powders synthesized depend on the experimental conditions. The average particle size of the powders, as determined using ImageJ software, is listed in **Table 3**.

Based on the FESEM microstructures and XRD patterns shown in **Figures 2** and **3**, it can be inferred that an increase in the citric acid to raw material ratio from 1 to 2 and 3 may result in increased heat release during combustion. This increased heat release leads to higher combustion reaction enthalpy, causing complete combustion reaction and particle size growth. Additionally, this phenomenon is connected to particle aggregation induced by the citrate. It seems that the presence of citrate promotes particle

Figure 2.
FESEM image of samples: (1) C0.2R2P0.5, (2) C1R3P3, (3) C0.2R1P1.5, (4) C0.6R2P1.5, (5) C0.2R3P0.5, (C0.6R1P3, (7) C1R3P1.5, (8) C1R2P1.5, and (9) C0.2R2P1.5 (Extracted from research done by the author [27]).

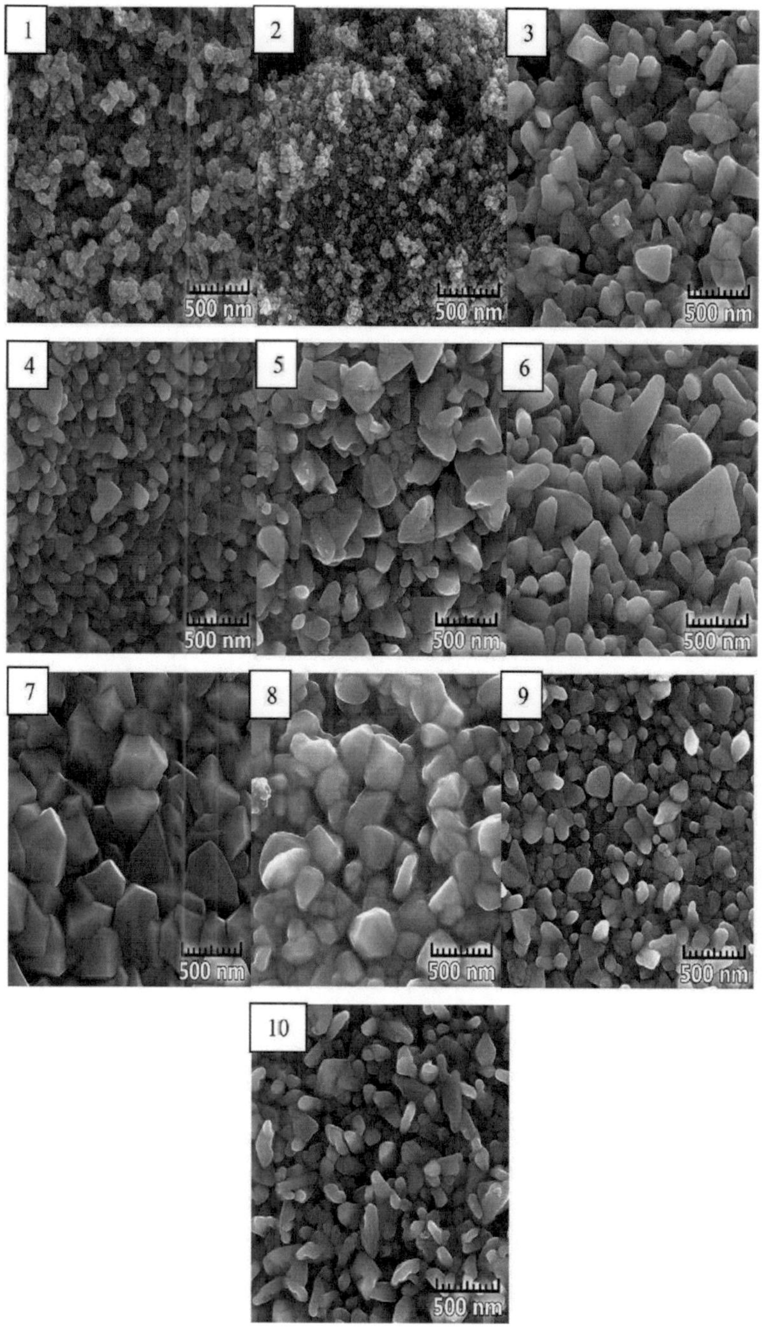

Figure 3.
FESEM image of samples: (1) C1R1P0.5, (2) C0.2R1P0.5, (3) C1R1P3, (4) C0.2R1P3, (5) C0.6R3P0.5, (6) C0.6R2P0.5, (7) C0.6R3P3, (8) C0.6R2P3, (9) C0.2R3P1.5, and (10) C1R2P0.5 (Extracted from research done by the author [27]).

aggregation by causing a build-up of charge on the surface, leading to particles sticking together to form larger particles. As the concentration of citric acid increased, SnO_2 particles changed from regular spheres to irregular shapes, and finally to polyhedron

Sample code	Particle size (nm)
$C_{0.2}R_2P_{0.5}$	80
$C_1R_3P_3$	310
$C_{0.2}R_1P_{1.5}$	40
$C_1R_2P_{0.5}$	180
$C_{0.6}R_2P_{1.5}$	170
$C_{0.2}R_3P_{0.5}$	105
$C_{0.6}R_1P_3$	132
$C_1R_3P_{1.5}$	270
$C_{0.2}R_3P_{1.5}$	125
$C_1R_2P_{1.5}$	208
$C_{0.2}R_2P_{1.5}$	100
$C_1R_1P_{0.5}$	55
$C_{0.2}R_1P_{0.5}$	20
$C_1R_1P_3$	175
$C_{0.2}R_1P_3$	75
$C_{0.6}R_3P_{0.5}$	225
$C_{0.6}R_2P_{0.5}$	148
$C_{0.6}R_3P_3$	245
$C_{0.6}R_2P_3$	215

Table 3.
Mean particle size of the FTO samples.

shapes. The formation of polyhedron shapes is believed to be a result of smaller particles self-assembling. The system's thermodynamics mediates the growth of self-assembly and is also time-dependent [18].

According to FESEM microstructures and XRD patterns in **Figures 1–3**, it can be concluded that increasing the concentration from 0.2 to 1 result in the growth of the particle size of powder.

Based on the microstructures observed using FESEM and the XRD patterns shown in **Figures 1–3**, it was found that when the pH is increased from 0.5 to 3, there is a significant rise in both the combustion reaction and the volume of external gases. This is attributed to the creation of an oxidizing environment around metallic ions due to the presence of nitrate ions. These nitrate ions contribute to the burning of the organic part of the gel, resulting in an increase in pH value and NO_{3-} ions. All these factors contribute to an increase in the combustion reaction rate and the growth of powder grains [27].

According to **Tables 2** and **3**, as the concentration, fuel ratio, and pH increase, the size of crystallites and particles also increases. The purpose of synthesizing nanopowders through test samples is to achieve the smallest particle size and identify the contributing conditions responsible for it. The optimal synthesized powder is sample $C_{0.2}R_1P_{0.5}$, with the lowest particle size of 20 nm. This sample was obtained using a precursor concentration of 0.2, a citric acid to precursor molar ratio of 1, and a pH value of 0.5. **Figure 4a–c** shows the XRD pattern, FESEM microstructure, and XPS of the C0.2R1P0.5 code, respectively. Photoelectron peaks for F 1 s, Sn 3d, O 1 s, and C

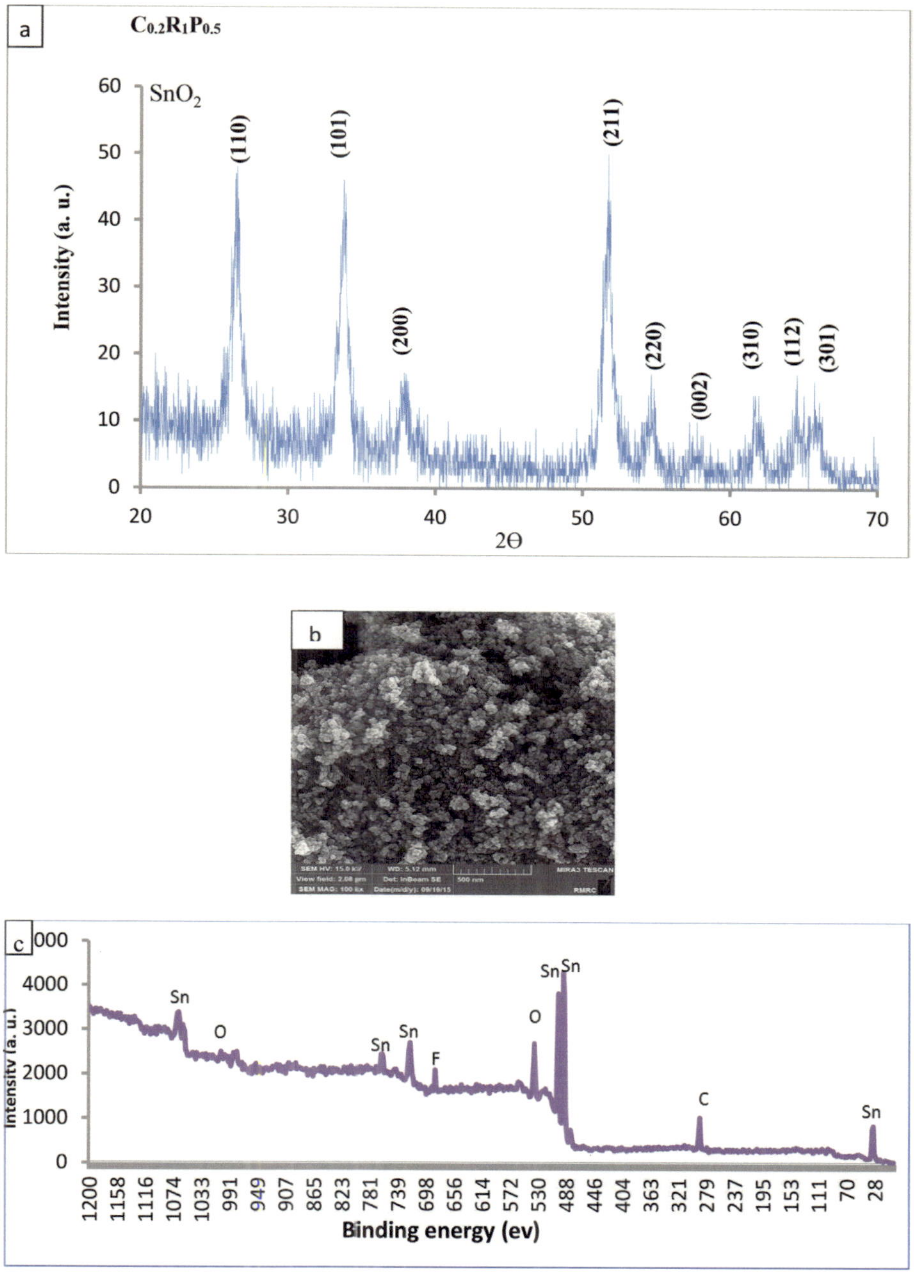

Figure 4.
(a) XRD pattern, (b) FESEM micrograph magnification (500 nm scale bar), and (c) XPS of $C_{0.2}R_1P_{0.5}$ code (Extracted from research done by the author [27]).

1 s were detected for the sample in the binding energy range of 0–1200 eV as shown in **Figure 4c**. The binding energies of the F 1 s, Sn 3d, and O 1 s photoelectron peak are at 690, 492, and 536 eV, respectively. A C 1 s peak at a binding energy of 290 eV is also observed in the sample.

By using $SnCl_4 \cdot 5H_2O$ as a precursor and citric acid as a fuel, the precursor mixture did not react at temperatures below 300°C. The precursor mixture reacted at 350°C.

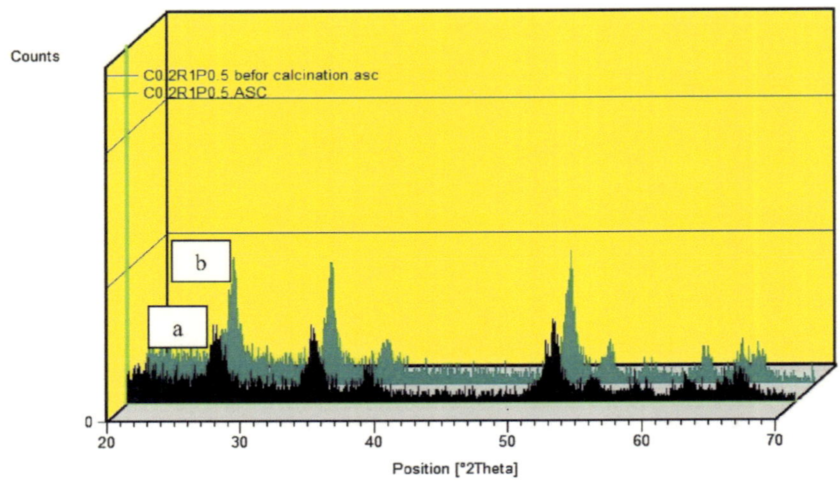

Figure 5.
XRD pattern of $C_{0.2}R_1P_{0.5}$ code: (a) after sol-gel combustion and (b) after calcination at 600°C (Extracted from research done by the author [27]).

From the XRD patterns in **Figure 5a** and **b** before calcination at 600°C, the crystalline phase was dominant and according to the width of the diffraction peak is composed of nano-size crystallite.

4. Conclusion

FTO was effectively produced through a gel combustion technique utilizing citric acid as a fuel and employing a D-Optimal factorial design. The size of the FTO nano-powder was most significantly influenced by the ratio of fuel to raw material, concentration, and pH value, respectively. The optimum synthesized powder had an average and a homogeneous size of 20–25 nm and a spherical morphology with a fuel to raw material ratio of 1, concentration value of 0.2, and pH value of 0.5. The most appropriate ratio for the fuel to raw material was found to be 1.

Author details

Sara Malek
Department of Engineering, North Tehran Branch, Islamic Azad University, Tehran, Iran

*Address all correspondence to: s_malek2000@yahoo.com

References

[1] Fortunato E, Ginley D, Hosono H, Paine DC. Transparent conducting oxides for photovoltaics. MRS Bulletin. 2007;**32**:242-247

[2] Hara K, Arakawa H, Luque A, Hegedus S. Handbook of Photovoltaic Science and Engineering. New York: John Wiley & Sons Ltd; 2003. p. 663

[3] Treharne RE, Durose K. Fluorine doped ZnO thin films by RF magnetron sputtering. Thin Solid Films. 2011;**519**(21):7579-7582

[4] Han CH, Han SD, Gwak J, Khatkar P. Synthesis of indium tin oxide (ITO) and fluorine-doped tin oxide (FTO) nano-powder by sol-gel combustion hybrid method. Materials Letters. 2007;**61**:1701-1703

[5] Aruna ST, Mukasyan AS. Combustion synthesis and nano materials. Current Opinion in Solid State and Materials Science. 2008;**12**:44-50

[6] Oshima M, Takemoto Y, Yoshino K. Optical and electrical characterization of FTO films grown by spray pyrolysis method. Physica Status Solidi C: Current Topics in Solid State Physics. 2009;**6**(5):1124-1126

[7] Girtan M, Bouteville A, Rusu GG. Preparation and properties of SnO_2: F thin films. Journal of Optoelectronics and Advanced Materials. 2006;**8**(1):27-30

[8] Geoffroy C, Campet G, Menil F, Portier J, Salardenne J, Couturier G. Optical and electrical properties of SnO_2: F thin films obtained by R.F. Sputtering with various targets. Active and Passive Electronic Components. 1991;**14**:111

[9] Maruyama T, Akagi H. Fluorine-doped tin dioxide thin films prepared by radio-frequency magnetron sputtering. Journal of the Electrochemical Society. 1996;**143**:283

[10] Stjerna B, Olsson E, Granqvist CG. Optical and electrical properties of radio frequency sputtered tin oxide films doped with oxygen vancancies, F, Sb, or Mo. Journal of Applied Physics. 1994;**76**:3797

[11] Chen PJ, Liao BH, Kuo CC, Lee CC. Improving the optical and electrical properties of fluorine-doped tin oxide films by various post-annealing treatments. Proceedings of SPIE. 2010;**77860Q**

[12] Ellmer K, Welzel T. Reactive magnetron sputtering of transparent conductive oxide thin films: Role of energetic particle (ion) bombardment. Journal of Materials Research. 2012;**27**:765

[13] Moure-Flores F d, Guillen-Cervantes A, Nieto-Zepeda KE, Quinones-Galvan JG, Hernandez-Hernandez A, Olvera M d l L, et al. f:F thin films deposited by RF magnetron sputtering: Effect of the SnF2 amount in the target on the physical properties, rev. Revista Mexicana de Fisica. 2013;**59**:335-338

[14] Seo M, Akutsu Y, Kagemoto H. Preparation and properties of Sb-doped SnO_2/metal substrates by sol-gel and dip coating. Ceramics International. 2007;**33**:625-629

[15] Fraigi L. Novel method to prepare nanocrystalline SnO_2 powders by a gel-combustion process. Nanostructured Materials. 1999;**11**(3):311-318

[16] Neri G. Ethanol sensors based on Pt-doped tin oxide nanopowders synthesized by gel-combustion. Sensors and Actuators, B. 2006;**117**:196-204

[17] Li Z, Shen W, Zhang X, Fang L, Zu X. Controllable growth of SnO_2 nanoparticles by citric acid assisted hydrothermal process. Colloids Surfaces A: Physicochemical and Engineering Aspects. 2008;**17**:327

[18] Sikhwivhilu LM, Pillai SK, Hillie TK. Influence of citric acid on SnO_2 nanoparticles synthesized by wet chemical processes. Journal of Nano science and Nanotechnology. 2011;**11**(6):4988-4994

[19] Adnan R, Razana N, Rahman IA, Farrukh MA. Synthesis and characterization of high surface area tin oxide nanoparticles via the sol-gel method as a catalyst for the hydrogenation of styrene. Journal of the Chinese Chemical Society. 2010;**57**:222-229

[20] Wu S, Yuan S, Shi L, Zhao Y, Fang J. Preparation, characterization and electrical properties of fluorine-doped tin dioxide nano crystals. Journal of Colloid and Interface Science. 2010;**346**:12-16

[21] Senthilkumar V, Vickraman P, Ravikumar R. Synthesis of fluorine doped tin oxide nanoparticles by sol-gel technique and their characterization. Journal of Sol-Gel Science and Technology. 2010;**53**(3):16-321

[22] Zhang J, Gao L. Synthesis of antimony-doped tin oxide (ATO) nanoparticles by the nitrate-citrate combustion method. Materials Research Bulletin. 2004;**39**:2249-2255

[23] Yu J, Duan LB, Wang YC, Rao GH. Influence of fuel-to-oxidizer ratio on the magnetic properties of Fe-doped. Journal of Solid State Chemistry. 2009;**182**:1563-1569

[24] Oliveiraa F, Varelab J, Orlandib M. Influence of pH in obtaining indium tin oxide nanoparticles by microwave assisted solvothermal method. Materials Research. 2018;**21**(2)

[25] Telford JK. A brief introduction to design of experiments. Johns Hopkins APL Technical Digest. 2007;**27**:224

[26] Antony J. Design of Experiments for Engineers and Scientists. Amsterdam: Elsevier Science & Technology Books; 2003

[27] Malek S, Baghshahi S, Sarraf-Mamoory R, Nemati A. Gel combustion synthesis of fluorine-doped tin oxide and its characteristics: Applying D-optimal factorial design of experiment. Bulletin of Materials Science. 2019;**42**:109

Chapter 5

Sol-Gel Synthesis of Hydroxyapatite: Applications, Methods, and a Case Study

Susana Devesa, Zohra Benzarti, Sílvia Soreto and Sandra Carvalho

Abstract

Hydroxyapatite (HAp) is a well-known biomaterial that, due to its biocompatibility and bioactivity, has been widely utilized in biomedical applications, including bone tissue engineering and drug delivery systems. Over the years, significant advancements have transformed HAp from a simple biocompatible substance into an advanced functional material with a wide range of applications. The synthesis of HAp is a complex area of study involving numerous techniques, each offering unique benefits and challenges. This chapter discusses different sol-gel synthesis routes, emphasizing the influence of precursor materials and solvent systems on the phase purity and morphology of the resulting materials. In addition to reviewing existing literature, this chapter presents a detailed case study on the synthesis of HAp and its dielectric properties. The case study addresses the challenges encountered during the process, emphasizing the necessity of carefully optimizing precursor concentrations to achieve the desired phase purity. The findings indicate that the highest content of hydroxyapatite (HAp) was obtained after heat treatment at 500°C, although secondary phases such as $Ca_2P_2O_7$, $Ca_3(PO_4)_2$, and Ca_3N_2 were also identified. The morphological analysis revealed particles of varying shapes and sizes with signs of agglomeration. Additionally, the electrical characterization showed that the grain and grain boundary resistance values were 46.61 and 245.63 MΩ, respectively. Through this combined review and case study approach, the chapter aims to provide a comprehensive overview of both the theoretical and practical aspects of HAp synthesis, offering valuable insights for researchers in the field of materials science.

Keywords: calcium orthophosphates, hydroxyapatite, sol-gel synthesis, precursors, impedance spectroscopy

1. Introduction

Calcium orthophosphates are chemical compounds that hold significant interest across various scientific disciplines, such as geology, chemistry, biology, and medicine, due to their natural abundance and presence in living organisms [1–4].

The subject of calcium orthophosphates has been under investigation for nearly 250 years [4]. At the end of the eighteenth century, the German geologist Abraham Werner coined the term "apatite" to describe a range of phosphate minerals that are now the major global source of phosphorus [5]. In the same year, Joseph-Louis Proust and Martin Klaproth proposed that calcium apatite (at the time, all calcium phosphates were referred to as apatites [1]) was the primary inorganic component of bones [5, 6]. The initial attempts to determine the chemical composition of calcium apatites and other calcium orthophosphates were made by Jacob Berzelius in the mid-nineteenth century [6–8]. Approximately 80 years later, H. Hausen proposed the existence of different calcium phosphate phases [1, 2]. These and other milestones in the historical development of the subject, as reported by Dorozhkin [6] (in his thorough investigation) and others [1, 2, 5], are depicted in **Figure 1**.

All calcium orthophosphates are composed of three key elements: calcium, with a + 2 oxidation state; phosphorus, with a + 5 oxidation state; and oxygen, with a − 2 reduction state, which form part of orthophosphate anions. Additionally, many calcium orthophosphates include hydrogen as part of acidic orthophosphate anions (e.g., HPO_4^{2-} or $H_2PO_4^-$), hydroxide ions (such as in $Ca_{10}(PO_4)_6(OH)_2$), and/or water molecules (e.g., $CaHPO_4{\cdot}2H_2O$). The chemical composition of these compounds can vary, producing a wide range of calcium phosphates distinguished by the type of phosphate anion involved—ortho-(PO_4^{3-}), meta-(PO_3^-), pyro-($P_2O_7^{4-}$), and poly-$(PO_3)_n^{n-}$. When multi-charged anions like orthophosphates and pyrophosphates are involved, calcium phosphates are also categorized by the number of hydrogen ions associated with the anion. Examples include mono-[$Ca(H_2PO_4)_2$], di-($CaHPO_4$), tri-[$Ca_3(PO_4)_2$], and tetra-($Ca_2P_2O_7$) calcium phosphates, with the prefixes "mono," "di," "tri," and "tetra" indicating the number of hydrogen ions replaced by calcium [6–8].

The atomic structure of calcium orthophosphates is centered around a network of orthophosphate (PO_4^{3-}) groups, which provides structural stability. Most calcium orthophosphates have low solubility in water but are readily soluble in acids while remaining insoluble in alkaline solutions. Chemically pure calcium orthophosphates are typically white crystals with moderate hardness. In contrast, natural calcium

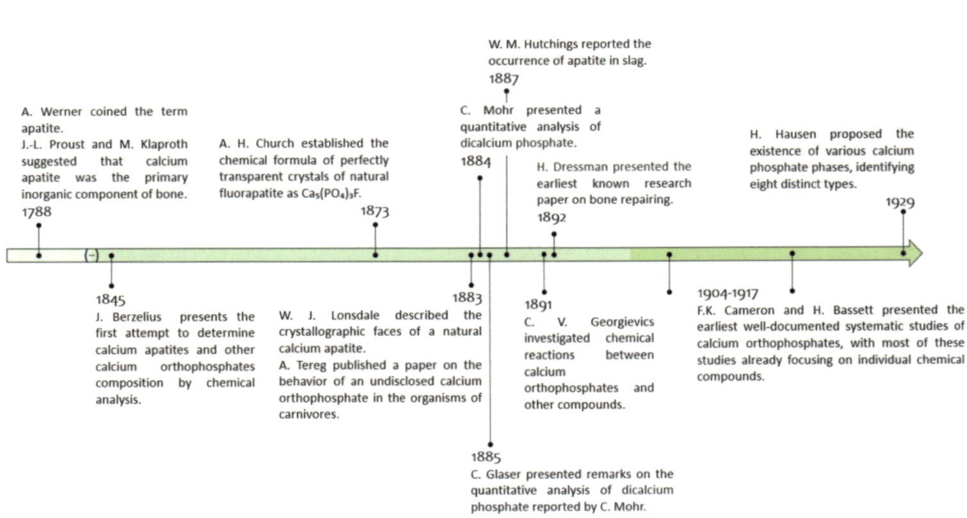

Figure 1.
Timeline of key milestones in the historical development of knowledge about calcium orthophosphates.

orthophosphate minerals are often colored due to the presence of impurities, particularly ions of iron, manganese, and rare earth elements [7, 8].

The names, chemical formulas, and Ca/P molar ratios of the known calcium orthophosphates are depicted in **Table 1**.

Among the various calcium orthophosphates, hydroxyapatite (HAp) has garnered significant attention due to its structural and chemical characteristics, as well as its stability within the pH range of 4–14 at room temperature [1]. These properties make it a compound of considerable interest in catalysis, the fertilizer industry, pharmaceutical products, protein chromatography applications, water treatment processes, and the preparation of biocompatible materials. Most notably, it is the main inorganic component in calcified hard tissues (for example, bone and teeth) of vertebrates [1, 2].

HAp has been extensively studied, surpassing other apatites such as fluorapatite and oxyapatite. This is evidenced by the analysis of papers published in the SCOPUS database over the past 10 years, as shown in **Figure 2**.

The number of published articles was retrieved from the SCOPUS database using the fields 'Article title, Abstract, Keywords' for the period from 2014 to 2024, with the keywords identified in the respective plots.

To emphasize the growing importance of hydroxyapatite research, **Figure 3** presents a comprehensive analysis of publication and citation trends (2000–2024), sourced from app.dimensions.ai, using the keyword "hydroxyapatite." Publications saw a steady upward trajectory, reflecting increased research activity and output. Citations also grew sharply, particularly in the later years, surpassing 1 million

Name	Chemical formula	Ca/P molar ratio
Monocalcium phosphate monohydrate	$Ca(H_2PO_4)_2{\cdot}H_2O$	0.5
Monocalcium phosphate anhydrous	$Ca(H_2PO_4)_2$	0.5
Dicalcium phosphate dihydrate (mineral brushite)	$CaHPO_4{\cdot}2H_2O$	1.0
Dicalcium phosphate anhydrous (mineral monetite)	$CaHPO_4$	1.0
Octacalcium phosphate	$Ca_8(HPO_4)_2(PO_4)_4{\cdot}5H_2O$	1.33
α-Tricalcium phosphate	α-$Ca_3(PO_4)_2$	1.5
β-Tricalcium phosphate	β-$Ca_3(PO_4)_2$	1.5
Amorphous calcium phosphates	$Ca_xH_y(PO_4)_z{\cdot}nH_2O$, n = 3–4.5; 15–20% H_2O	1.2–2.2
Calcium-deficient hydroxyapatite	$Ca_{10-x}(HPO_4)_x(PO_4)_{6-x}(OH)_{2-x}$ $(0 < x < 1)$	1.5–1.67
Hydroxyapatite	$Ca_{10}(PO_4)_6(OH)_2$	1.67
Fluorapatite	$Ca_{10}(PO_4)_6F_2$	1.67
Oxyapatite	$Ca_{10}(PO_4)_6O$	1.67
Tetracalcium phosphate (mineral hilgenstockite)	$Ca_4(PO_4)_2O$	2.0

Table 1.
Existing calcium orthophosphates: names, chemical formulas, and Ca/P molar ratios [7, 9].

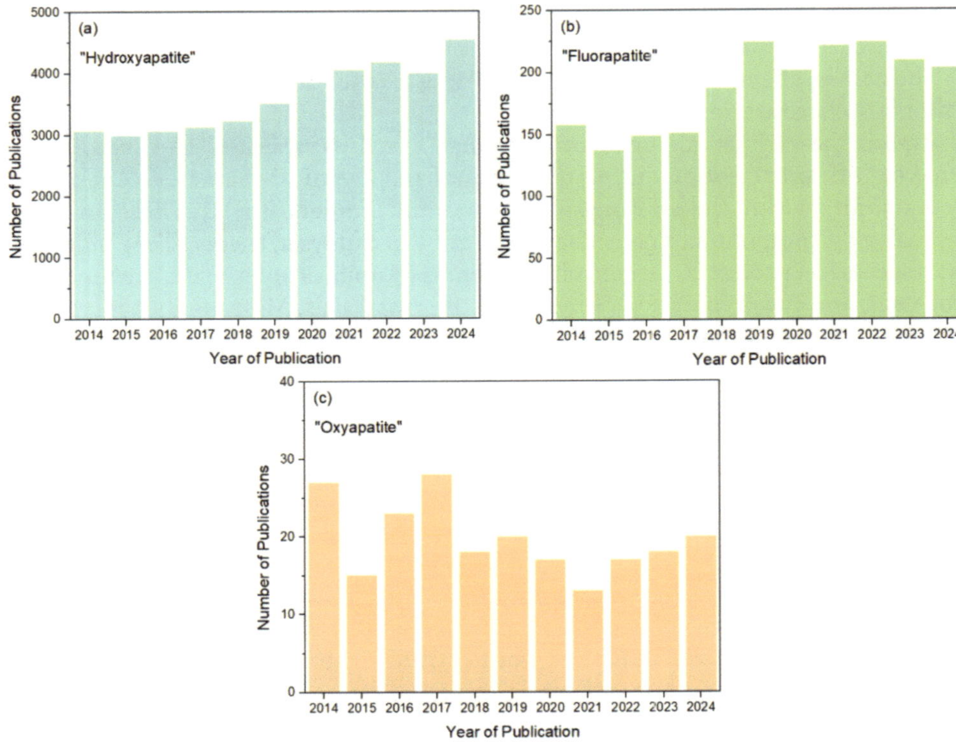

Figure 2.
Number of published articles retrieved using the specified keywords (SCOPUS database): (a) "Hydroxyapatite"; (b) "Fluorapatite"; (c) "Oxyapatite."

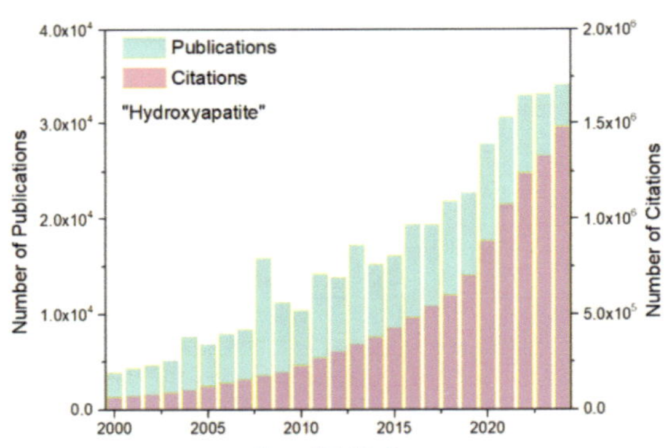

Figure 3.
Number of published articles and citations retrieved using the keyword "hydroxyapatite" (app.dimensions.ai), from 2000 to 2024.

annually, which suggests the field has gained considerable academic and industrial impact. The parallel growth of citations with publications highlights the increasing recognition and influence of hydroxyapatite research during this period. Therefore, it is imperative to ensure that researchers have access to the latest information and advancements.

HAp has the general formula $Ca_5(PO_4)_3(OH)$ but is usually written as $Ca_{10}(PO_4)_6(OH)_2$ to emphasize that the crystal unit cell comprises two molecules [10].

Chemically pure HAp crystallizes in the monoclinic space group $P2_1/b$, as suggested in 1967 and later confirmed by single-crystal structure analysis [11–13], with cell parameters a = 9.4214 [8] Å, b = 2a, c = 6.8814 [7] Å, and $\alpha = \beta = 90°$, $\gamma = 120°$ [14]. The most common and initially known form is hexagonal, described in the space group $P6_3/m$ (N° 176) with lattice parameters a = b = 9.432 Å, c = 6.881 Å, and $\alpha = \beta = 90°$, $\gamma = 120°$ [13]. The phase transition from monoclinic to hexagonal in hydroxyapatite occurs at approximately 470 K [12].

The unit cell of hexagonal HAp, comprising 44 atoms [15], is depicted in **Figure 4**. The 10 calcium ions (4 Ca_I and 6 Ca_{II}) occupy two distinct sites. The four Ca_I ions are coordinated by nine oxygen atoms from phosphate tetrahedra, while the six Ca_{II} ions are coordinated by six oxygen atoms from five phosphate groups and one hydroxyl ion. The Ca_I sites form distinct calcium channels parallel to the c-axis, whereas the Ca_{II} ions create triangular bases staggered by 60° above and below the hydroxyl ions. Similar to Ca_I, the hydroxyl ions form distinct channels parallel to the c-axis. Based on their coordination with oxygen and calcium, the oxygen atoms are categorized into three types: O_I, O_{II}, and O_{III} [13, 15, 17–19].

HAp structure allows other groups, anionic and cationic substances, to replace Ca^{2+}, PO_4^{3-}, and OH^- ions, which is considered a unique attribute since these modifications enable the tailoring of HAp for specific applications. These substitutions can affect network parameters, crystallinity, crystal dimensions, surface texture, solubility, spectral properties, and thermal stability [3, 20].

HAp can be derived from both natural and synthetic sources. Natural sources highlighted in the literature include biological materials or wastes such as mammalian bones (for example, bovine, camel, and horse), marine or aquatic sources (like fish bones and fish scales), shell sources (including cockle, clam, eggshell, and seashell), plants, algae, and mineral sources such as limestone [3, 21]. However, natural hydroxyapatite (HAp) is non-stoichiometric and may be deficient in either calcium or phosphorus. Regardless of its origin, HAp often contains trace impurities such as phosphite ions (PO_3^{3-}), chloride ions (Cl^-), and fluoride ions (F^-). Notably, PO_3^{3-} and Cl^- have been reported to weaken the HAp structure, whereas F^- is known to enhance its strength [1].

As mentioned before, HAp is the main component of bones and teeth. It is highly biocompatible and can be a very cost-effective material due to its high availability.

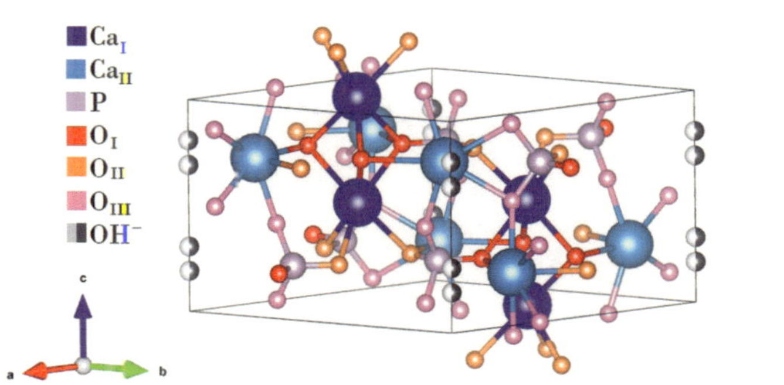

Figure 4.
The unit cell of hexagonal HAp (drawn using VESTA software, based on atom coordinates and site occupations from reference [16]).

This, combined with the evolution of biotechnology in recent years, has increased the interest in nature-inspired materials while creating new applications beyond the initial scope. These facts have made HAp one of the most attractive and important bioceramics [22].

2. Applications

HAp has long been valued for its outstanding biocompatibility and bioactivity, establishing itself as a key material in bone tissue engineering, orthopedic implants, and dental applications [23, 24].

Furthermore, it has osteoconductive, non-toxic, non-inflammatory, and non-immunogenic properties [25].

Its capacity to bond with surrounding tissues, enhance cellular attachment, and support osseointegration has positioned it as a preferred choice for biomedical implants and surface coatings. Furthermore, more recent research indicates that HAp particles may play a role in suppressing the growth of various cancer cell types [23, 24].

As knowledge about HAp expanded, its applications evolved beyond conventional biomaterials [23]. Advances in synthesis techniques and material engineering revealed unique properties of HAp that broadened its scope to various fields. By modifying its composition, morphology, and surface features, researchers achieved improvements in mechanical strength, drug delivery efficiency, and biodegradability [23, 24]. These adjustments expanded the use of HAp to include drug delivery systems, biosensors, tissue engineering scaffolds, and applications in regenerative medicine [25–28].

In addition, HAp's outstanding biomineralization capabilities have facilitated the integration of functional ions and molecules during its synthesis. This has paved the

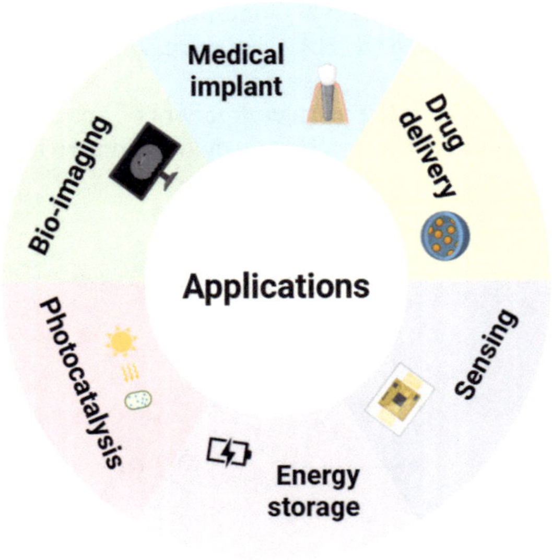

Figure 5.
Schematic representation of the multiple applications of HAp (Created in BioRender. https://BioRender.com/z10x819).

way for creating bioactive coatings and composites with targeted therapeutic properties. Functionalized HAp materials have shown significant potential in antimicrobial coatings, controlled release systems for growth factors and drugs, catalysts in various chemical processes, sensing, and energy storage, demonstrating their potential as advanced functional materials beyond traditional biomedical applications [23, 29–34].

Recently, HAp nanoparticles and nanostructured materials have gained significant attention in research due to their distinctive physicochemical characteristics and their potential in targeted drug delivery, imaging, and theranostic applications [23, 35–37]. The capacity to control the size, shape, and surface properties of HAp at the nanoscale has opened new possibilities for innovative strategies in personalized medicine and regenerative therapies [23, 38].

Figure 5 schematically shows some of the possible applications of HAp, highlighting its role as an advanced multifunctional material.

3. Synthesis methods

HAp, as an advanced material in biomedical applications, is synthesized using various methods to optimize its structural, physicochemical, and biological properties. The synthesis methodologies can broadly be categorized into dry methods, wet methods, high-temperature methods, and hybrid techniques. Each method offers unique advantages and limitations depending on the desired application and characteristics of the HAp product.

A brief description of the most common methods is presented below and schematically summarized in **Figure 6**.

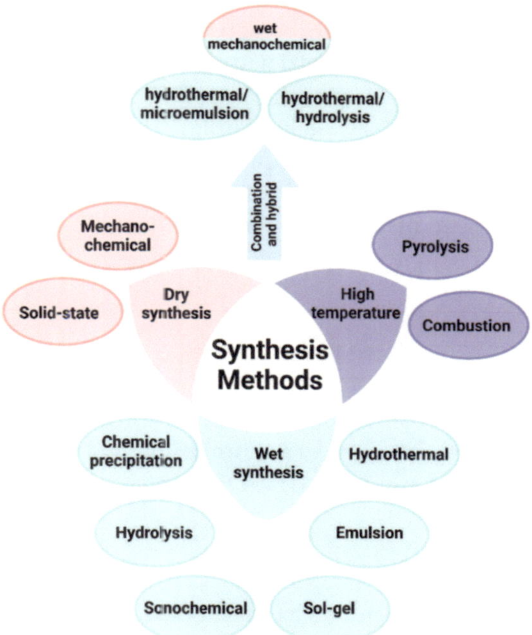

Figure 6.
Example of techniques used to synthesize HAp [Created in BioRender. https://BioRender.com/k64r573].

3.1 Dry synthesis methods

Solvent-free dry methods are characterized by their simplicity and low cost, which makes them suitable for large-scale manufacturing [39]. However, these methods generally produce HAp particles with larger sizes and less defined shapes compared to wet chemical methods. These techniques generally involve high-temperature processing to induce the formation of HAp from calcium and phosphate precursors [40]. Among the dry methods, solid-state synthesis and mechanochemical synthesis are the most prominent.

3.1.1 Solid-state method

The solid-state synthesis method entails mixing calcium and phosphate precursors in stoichiometric ratios (Ca^{2+}/PO^{3-}), followed by calcination at temperatures above 700°C [41]. This high-temperature treatment promotes the formation of highly crystalline HAp particles. However, the process is limited by the low ionic diffusion coefficients within the solid phase, leading to heterogeneity in the phase composition and irregular particle shapes. To overcome these limitations, molten salt synthesis (MSS) has been developed as a variant of the solid-state method.

In MSS, a molten salt medium facilitates the dissolution and subsequent reaction of the solid precursors [42] and enables the formation of products at lower temperatures [43]. This technique has been particularly effective in producing well-defined HAp structures, including whiskers and rods, under controlled conditions.

3.1.2 Mechanochemical synthesis

Mechanochemical synthesis, also known as ball milling, involves the application of mechanical energy to induce chemical reactions and structural transformations in the precursors [44]. This method is efficient and reproducible, offering a cost-effective approach to producing HAp with controlled particle size and morphology. The primary processing parameters, such as milling time, ball-to-powder ratio, and rotation speed, significantly influence the properties of the synthesized HAp [45]. Extended milling times have been shown to reduce particle size and enhance lattice strain, resulting in improved material properties [39].

3.2 Wet synthesis methods

Wet synthesis methods are favored for producing HAp nanoparticles with uniform morphology and high purity [46]. These techniques utilize aqueous solutions as reaction media, enabling precise control over the nucleation and growth processes. Key wet synthesis methods include chemical precipitation, hydrolysis, sol-gel synthesis, hydrothermal synthesis, and emulsion-based techniques.

3.2.1 Chemical precipitation

Chemical precipitation is the most straightforward wet method, involving the reaction of calcium and phosphate salts in aqueous solutions [47]. The resulting suspension is either directly processed into a powder or aged at atmospheric pressure before filtration, drying, and further processing [48]. The neutralization reaction that occurs upon mixing the $Ca(OH)_2$ suspension and the phosphoric acid solution can be described by the following equation [49]:

$$10Ca(OH)_2\,(\text{susp.}) + 6H_3(PO_4)(\text{sol.}) \rightarrow Ca_{10}(PO_4)_6(OH)_2\,(s) + 18H_2O \quad (1)$$

The process typically requires maintaining a pH above 4.2 to ensure the formation of HAp over other calcium phosphate phases. The reaction conditions, including temperature, pH, and reactant concentration, are critical in determining the crystallinity and stoichiometry of the resulting HAp. Despite its simplicity, this method often yields HAp with low crystallinity, necessitating post-synthesis heat treatments to enhance material properties.

3.2.2 Hydrolysis

Hydrolysis involves the transformation of calcium phosphate precursors, such as tricalcium phosphate, into HAp under specific conditions [50]. The process is typically conducted at elevated temperatures and controlled pH levels to optimize the Ca/P ratio and promote complete conversion. Dual-stage hydrolysis methods allow for separate control of the initial transformation and subsequent compositional adjustments, yielding stoichiometric and crystalline HAp. First, dicalcium phosphate dihydrate rapidly converts to non-stoichiometric HAp at elevated temperatures (40–80°C). In the second stage, the Ca/P ratio is adjusted to 1.67 by pH modification and Ca^{2+} addition. The synthesized HAp powders exhibited a porous structure composed of dense aggregates of non-uniform, thin microcrystals and a relatively low crystallinity [51].

3.2.3 Hydrothermal synthesis

Hydrothermal synthesis utilizes high pressure and temperature in a sealed autoclave to produce HAp directly from solutions [52]. Increasing hydrothermal treatment temperature or pressure generally leads to an improvement in the Ca/P ratio of the precipitates [53]. This method promotes the formation of highly crystalline HAp with minimal impurities. Parameters such as reaction temperature, pressure, and the addition of surfactants influence the size and morphology of the final product. Hydrothermal treatment is especially suitable for producing well-ordered nanostructures and has applications in fabricating HAp coatings.

3.2.4 Sol-gel method

The sol-gel technique involves the formation of a colloidal solution (sol), which is subsequently gelled to create a network that is calcined to produce HAp. This method allows molecular-level mixing of calcium and phosphorus precursors, ensuring uniform composition and high purity [54]. This method, with its precise control over process parameters, leads to improved HAp composition and enhanced biocompatibility in vivo. The sol-gel process is flexible, enabling the incorporation of dopants to modify HAp properties [55]. Challenges include the need for precise control over reaction conditions to avoid phase impurities [56].

3.2.5 Emulsion method

The emulsion method employs immiscible liquids to create microenvironments where HAp precursors react. This technique effectively controls particle size and morphology, reducing agglomeration [57]. The emulsion method offers several advantages, including its facile synthesis route that avoids the need for high temperatures in the initial stages. This approach is particularly well-suited for the production of nanostructured HAp with narrow size distributions [58–60]. However, the use of organic solvents poses environmental and safety concerns.

3.2.6 Sonochemical synthesis

Sonochemical methods utilize ultrasonic waves to induce cavitation within the solution, enhancing reaction kinetics and facilitating the formation of nanoscale HAp particles [61, 62]. This approach leverages the localized high temperatures and pressures generated during bubble collapse to accelerate chemical reactions involving liquid and solid reactants [63]. This approach produces highly uniform particles with minimal agglomeration [64, 65]. The increased surface area of sonochemically synthesized HAp makes it advantageous for applications requiring enhanced bioactivity.

3.3 High-temperature methods

High-temperature methods utilize thermal energy to synthesize or refine HAp particles [66]. These methods are efficient for producing highly crystalline HAp but may introduce secondary phases or structural imperfections.

3.3.1 Pyrolysis

The pyrolysis method, also known as 'spray pyrolysis,' for preparing HAp was developed within the field of bio-inspired chemistry and involves spraying precursor solutions into a flame or the hot zone of an electric furnace using an ultrasonic generator [45, 67]. It should be noted that the pyrolysis method can also be classified under the broader category of aerosol methods (or gas-phase methods), in which gas-to-particle or liquid-to-particle conversions occur during aerosol decomposition [21].

Alam et al. [68] successfully applied the spray pyrolysis method with calcium phosphate, calcium nitrate tetrahydrate, and diammonium hydrogen phosphate as precursors. The reactants were mixed at 1100°C with a carrier gas to produce the final powder, resulting in HAp powder composed of micron-sized spheres.

Compared to the combustion method, the pyrolysis synthesis does not require mixing fuel with reactants and can be easily scaled up for continuous HAp particle production [21, 69].

3.3.2 Combustion synthesis

The combustion method is a promising approach for synthesizing HAp, characterized by rapid, self-sustaining redox reactions between a fuel (e.g., citric acid, urea) and an oxidant in solution [70, 71]. This method typically involves a rapid temperature increase followed by rapid cooling, which promotes nucleation and restricts particle growth. Indeed, it offers advantages such as rapid synthesis, high purity, and the ability to control particle size and morphology [72]. However, careful optimization of reaction parameters, including fuel type, heating rate, and cooling rate, is crucial to minimize the formation of secondary phases like calcium oxide (CaO) and achieve high-purity HAp [73].

3.4 Combination and hybrid techniques

Hybrid synthesis approaches, such as the wet mechanochemical method, combine different techniques to enhance HAp production. These methods often influence the advantages of individual techniques while mitigating their limitations. For instance, the wet mechanochemical approach accelerates reaction kinetics compared to traditional mechanochemical methods by incorporating an aqueous environment. This hybrid strategy offers a significant advantage by eliminating the need for high-energy input typically associated with hydrothermal techniques while also circumventing the requirement for external drying or the use of a pressure vessel.

Other notable hybrid methods include hydrothermal/hydrolysis and hydrothermal/microemulsion techniques. These approaches offer distinct advantages, such as improved crystallinity (hydrothermal/hydrolysis) or the ability to produce small, narrowly distributed HAp particles (microemulsion). Moreover, innovative techniques like the use of ultrasound and microwave irradiation have been explored to further optimize HAp synthesis. The solvothermal method effectively synthesizes HAp rod particles with high crystallinity, minimal agglomeration, and narrow size distribution. This approach utilizes *Pomacea sp.* shell powder as a calcium source and citric acid as a complexing agent. To ensure a Ca/P ratio of 1.67, the phosphate precursor is carefully added to the shell powder [74]. The synthesis of calcium carbonate involves the reaction between poly(styrene sulfonic acid) sodium salt and calcium chloride. Subsequent microwave irradiation under alkaline conditions leads to the formation of HAp. Ramakrishnan et al. [75] demonstrated the role of wheat gluten in accelerating the thermal cross-linking of silica-HAp gelatine composites.

4. Sol-gel synthesis

The sol-gel synthesis method is a well-known and established process, valued for its flexibility and applicability to a diverse range of structural materials, with particle sizes spanning from the nanometer to micrometer scale [76].

One of the earliest proposed methods for the synthesis of HAp was indeed the sol-gel method [77]. However, only a limited number of studies have specifically investigated the synthesis of HAp using this process [77, 78]. Despite not being one of the most commonly applied methods for HAp synthesis, its application over the last 10 years has remained stable, as shown in **Figure** 7.

The data on published articles was obtained from the SCOPUS database by searching the fields "Article Title, Abstract, Keywords," from 2014 to 2024, using the terms "hydroxyapatite" and "sol-gel."

The sol-gel method is an effective approach for synthesizing nanophasic hydroxyapatite (HAp) due to its ability to provide strict control over process parameters [79]. This method enables molecular-level mixing of calcium and phosphorus, significantly enhancing the chemical homogeneity of HAp [79, 80].

Additional key benefits of the sol-gel process include the formation and fusion of arranged crystals at low temperatures, resulting in lower sintering temperatures compared to other methods. Moreover, the resulting powder exhibits a stoichiometric structure, a large surface area, and a small cluster size [77].

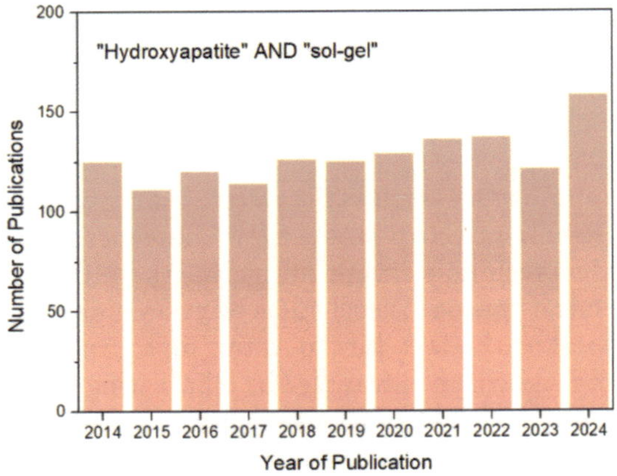

Figure 7.
Number of published articles retrieved using the specified keywords (SCOPUS database): "Hydroxyapatite" and "sol-gel."

It has also been reported that HAp materials synthesized via the sol-gel process are effective in improving contact and stability at artificial/natural bone interfaces in both in vitro and in vivo environments [77, 78].

However, despite these recognized advantages, there are also drawbacks to synthesizing HAp through the sol-gel method, namely the formation of a secondary phase, commonly calcium oxide (CaO), and the high cost of certain precursors, particularly alkoxide-based compounds [77, 81].

The sol-gel process essentially involves the synthesis of an inorganic network by mixing calcium and phosphorus precursors in solution, followed by hydrolysis, gelation, and calcination [80].

It has been proven that the calcination temperature—i.e., the required temperature to form the HAp structure—depends largely on the chemical nature of the precursors [82]. Regarding the solvent, Nazeer et al. [76] investigated the effect of three different solvents on the physical and chemical properties of hydroxyapatite synthesized via the sol-gel method. The authors demonstrated that the dielectric constant and polarity of the reaction solvent had a critical impact on the nature of the product obtained during synthesis. The solvent can influence the reaction rate and hydrolysis, ultimately affecting the structure of the synthesized material. According to the authors, solvents with high dielectric constants favor the complete dissociation of electrolytes, while solvents with low dielectric constants promote ion pairing.

Figure 8 schematically presents the typical sol-gel route, highlighting some of the precursors and solvents that have been applied over the years. However, it is worth noting that, according to Livage et al. [83], phosphate esters and phosphoric acid are not suitable for the sol-gel synthesis of phosphate materials, since the first cannot be easily hydrolyzed, while the second reacts too quickly, leading to precipitation rather than gelation. The same authors showed that more convenient precursors could be obtained by dissolving phosphorus pentoxide in alcohols. Nevertheless, this is not the only possibility for synthesizing phosphates from molecular precursors. Alkyl phosphites, as well as phosphoryl chloride, can also be used as molecular precursors.

Liu et al. [82] explored two different sol-gel routes for the synthesis of hydroxyapatite (HAp): one ethanol-based and the other water-based, both utilizing the same precursors—triethyl phosphite, $C_6H_{15}O_3P$, and calcium nitrate, $CaNO_3$.

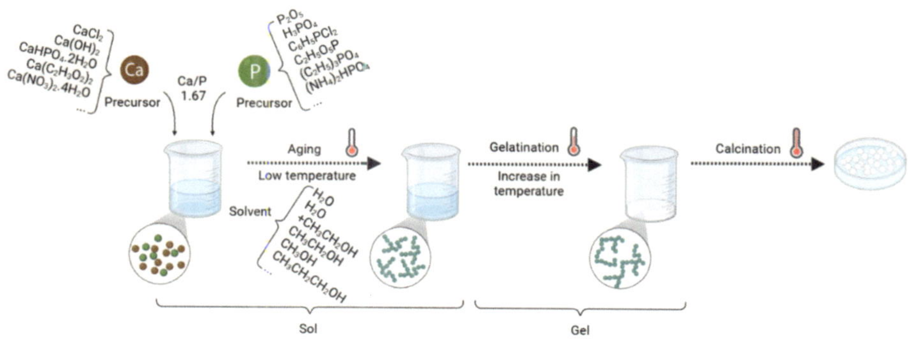

Figure 8.
Typical sol-gel route for the synthesis of HAp (Created in BioRender. https://BioRender.com/c54x906).

According to the authors, in anhydrous ethanol, alkyl groups may form $Ca(OR)_y(NO_3)_{2-y}$, replacing some of the nitrate groups. In contrast, calcium nitrate dissolves in water to form ionic species. Both ethanol- and aqueous-based sols were found to be relatively stable at room temperature, not forming a solid gel for at least 5 days.

After aging, the hydrolyzed phosphorus sol—which the authors hypothesized to be in the form of a phosphoric ester, $PHO(OEt)_2$, or more generally, $P(OEt)_{3-x}(OH)_x$ — interacted with the calcium sol. This interaction likely involved $Ca(OEt)_{3-x}(NO3)_{2-y}$ in anhydrous ethanol and Ca^{2+} ions in water, forming oligomeric derivatives containing Ca–O–P bonds.

As heating continues, solvents are eliminated, accompanied by increased thermal dehydration or the polymerization/condensation of the derivative units. This process results in the formation of additional Ca-O-P bonds in the dry gels.

As a result of the two routes investigated, the ethanol-based synthesis yielded a thermally stable calcium hydroxyapatite phase, whereas the water-based synthesis produced a calcium-deficient apatite.

5. Short review of hydroxyapatite synthesized by the sol-gel method

HAp is typically used in powdered form, and its efficacy is influenced by various properties, such as phase purity, crystallinity, precursor sources, solvent or surfactant materials, particle size, morphology, and surface area [23, 38].

A wide variety of routes have been adopted for the synthesis of HAp using the sol-gel method. Efforts have been made to control geometry, crystallinity, size, stoichiometry, and degree of particle agglomeration (for different applications) by employing new routes or modifying pre-existing synthesis methods [23].

This section provides a review, also summarized in **Table 2**, though not exhaustive, of some of these routes, focusing on the combination of different precursors and solvents, as well as on the phase purity and morphology of the obtained material.

Balamurugan et al. [79] used triethyl phosphite, dissolved in a mixture of ethanol and distilled water, and a stoichiometric amount of calcium nitrate tetrahydrate, according to the Ca/P ratio of 1.67, which was dissolved in ethanol and added dropwise to the first solution. This procedure was performed under vigorous stirring, which was continued for an additional 10 minutes after the mixing. As a result of this

Reference	Ca precursor	P precursor	Solvent	Complexing agents	Calcination temperature (°C)	Secondary phases	Morphological features
Balamurugan et al. [79]	Calcium nitrate tetrahydrate	Triethyl phosphite	Water/ ethanol	—	900	—	Nano-structured, ellipse-like
Bogdanoviciene et al. [84]	Calcium acetate monohydrate	Ammonium hydrogen phosphate	Water/ ethanol	Ethylene diamine tetra-acetic acid and tartaric acid	1000	—	Agglomerates ranging from 500 to 8 μm and elongated particles with lengths of about 2–10 μm and widths of 1–4 μm.
Padmanabhan et al. [85]	Calcium nitrate tetrahydrate	Potassium dihydrogen phosphate	Water	—	700	—	Nanorods of 70–90 nm diameter and 400–500 nm length
Chen et al. [86]	Calcium nitrate tetrahydrate	Trimethyl phosphite	Water/ ethanol	—	600	—	Clusters of numerous distinguishable grains sized approximately between 90 and 150 nm.
Figueroa et al. [87]	Tetrahydrate calcium nitrate	Pentoxide phosphorous	Ethanol	—	1200	Calcium phosphate	Average grain size of 500 nm
Bakan et al. [88]	Calcium nitrate tetrahydrate	Ammonium dihydrogen phosphate	Water/ ethanol	—	750	—	Nano-needles of 12 nm average diameter and 65 nm average length
Türk et al. [89]	Calcium hydroxide	Phosphoric acid	Water and ethanol	—	950	Calcium oxide and calcium hydroxide	Nano-sized spherical aggregates
Farazin et al. [90]	Calcium nitrate tetrahydrate	Diammonium hydrogen phosphate	Water/ ethanol	—	700	Beta-tricalcium phosphate and calcium phosphate	Average grain size of 48 nm

Table 2.
Summary of the referenced works, including precursor materials, solvent, heat-treatment temperature, and corresponding outcomes in terms of phase purity and morphology.

process, a clear solution was obtained and aged at room temperature for 16 hours. The mixture was then dried at 60°C until a viscous liquid was obtained, and further drying of the viscous liquid at 60°C resulted in a white gel. The gel was ground with a mortar and pestle into a fine powder and subjected to different calcination treatments, ranging from 300 to 900°C, for 2 hours.

According to the authors, a pure monophasic crystalline HAp phase was obtained with the calcination performed at 900°C, exhibiting a nano-structured, ellipse-like morphology.

In the sol-gel process presented by Bogdanoviciene et al. [84], ammonium hydrogen phosphate and calcium acetate monohydrate were selected as P and Ca precursors, respectively, with a Ca/P molar ratio of 1.67. Calcium acetate was first dissolved in ethanol at 65°C, and to this solution, ammonium hydrogen phosphate, dissolved in distilled water, was added. The resulting mixture was stirred for 1 hour at the same temperature. In a subsequent step, ethylene diamine tetra-acetic acid or tartaric acid, as complexing agents, were added to the solution. After concentrating the solution by slow evaporation at 65°C under stirring, a procedure that took 15 hours, the sol turned into a transparent gel, which was dried at 100°C. The obtained material was ground and annealed for 5 hours at 1000°C in air, with a heating rate of 10°C/min.

According to the X-ray Diffraction (XRD) analysis, regardless of the complexing agent used, no impurities were detected. Regarding the morphology, Scanning Electron Microscopy (SEM) micrographs of the HAp powders derived from the ethylene diamine tetra-acetic acid route suggest that the sample is composed of agglomerates with a broad size distribution, ranging from 500 to 8 μm. The HAp powders derived from the tartaric acid route present a different morphology, characterized by well-shaped ultrafine elongated particles (microrods and/or microsticks) with lengths of about 2–10 μm and widths of 1–4 μm.

For the synthesis of HAp nanorods, Padmanabhan et al. [85] used calcium nitrate tetrahydrate and potassium dihydrogen phosphate as calcium and phosphorus precursors, respectively, both dissolved in deionized water, with concentrations adjusted to ensure a Ca/P ratio of 1.67. This synthesis was conducted at pH = 9, using ammonia solution. The resulting mixture was stirred for 1 hour and aged at room temperature for 48 hours. After repeated procedures of filtration and washing with double-distilled water to remove $NH4^+$ and NO_3^- ions, the obtained precipitate was dried at 60°C for 24 hours in a dry oven. The dried powder was subsequently heated over a temperature range of 300–700°C for 30 minutes in an electric furnace in air, with a heating rate of 10°C/min.

XRD analysis showed that the sample heat-treated at 700°C exhibited a high degree of purity, with no secondary phases detected. Regarding the morphology, the authors reported the formation of nanorods with diameters of 70–90 nm and lengths of 400–500 nm, which has been attributed to the "oriented attachment" mechanism, where HAp crystallites adhere to high-energy planes and template growth in a specific direction.

A simple but time-consuming route using trimethyl phosphite and calcium nitrate tetrahydrate as precursors was presented by Chen et al. [86]. These were first dissolved in absolute ethanol, with deionized water added to the trimethyl phosphite solution under vigorous stirring. The solutions were mixed with a Ca/P molar ratio of 1.4, 1.5, and 1.67 (S1.4, S1.5, and S1.67, respectively). The prepared mixtures were continuously stirred for about 10 minutes and then kept at room temperature for approximately 16 hours. Afterwards, they were heated in a water bath at 60°C for 6 days, resulting in a white dried gel. The dried gel was then calcined in a furnace at 600°C for 3 hours.

The XRD analysis showed that sample S1.4 was composed of beta-tricalcium phosphate, sample S1.5 was a mixture of beta-tricalcium phosphate and HAp, with HAp being the main phase, while sample S1.67 was pure HAp.

The SEM micrographs revealed a similar morphology among the synthesized powders, showing essentially a mixture of particles of different sizes, with each particle being an assembly of numerous distinguishable grains sized approximately between 90 and 150 nm.

Figueroa et al. [87] proposed a considerably faster approach, using tetrahydrated calcium nitrate and phosphorus pentoxide as precursors, combined in a stoichiometric ratio of Ca/P = 1.67. The starting solutions were prepared separately, using ethanol as the solvent, and the first solution was added dropwise to the second solution.

The final mixture was kept under stirring for an hour at room temperature and then dried at 56°C for 12 hours. Finally, the obtained powder was ground and heat-treated at 600, 900, and 1200°C for 2 hours.

XRD analysis confirmed the formation of a crystalline HAp phase; however, a small trace of calcium phosphate was also identified.

Transmission electron microscopy (TEM) analysis revealed particles with relatively homogeneous shapes, averaging 40 nm in size for the heat-treatment temperature of 600°C, and 500 nm for 900 and 1200°C, indicating that the heat-treatment temperature may control particle size.

To explore the influence of aging time on the synthesis of HAp, Bakan et al. [88] used calcium nitrate tetrahydrate and ammonium dihydrogen phosphate as starting Ca and P precursors, respectively, maintaining a Ca/P ratio of 1.67. The first precursor was dissolved in ethanol and the second in water, with ammonia (NH_3, Merck) used to adjust the pH of the solution. The resulting mixture was kept under vigorous stirring at room temperature until a white precipitate was obtained, which was subsequently aged for 12, 24, and 48 hours, also at room temperature.

The gel obtained after the aging process was first filtered and washed repeatedly with double-distilled water to remove NH_4^+ and NO_3^- ions, and then washed with pure acetone. The obtained material was dried at 80°C for 12 hours in an oven, and the dried powder was calcined at 750°C for 4 hours in an electric furnace, with a heating rate of 10°C/min in air.

The XRD studies revealed that the samples synthesized with 12-hour and 24-hour aging times contained a small amount of beta-tricalcium phosphate along with HAp. After 48 hours of aging, the secondary phase gradually disappeared, and pure HAp was obtained.

From the SEM and TEM micrographs, the HAp particles were identified as hexagonal, nano-needle-like structures with a mean particle diameter of 12 nm and a mean particle length of 65 nm.

To study the influence of the solvent nature and calcination time, Türk et al. [89] synthesized HAp using calcium hydroxide and phosphoric acid, considering a Ca/P atomic ratio of 1.67. Two separate solutions of calcium hydroxide were prepared: one with double-distilled water and the other with ethanol. For the preparation of the phosphorous solutions, the same procedure was followed. The prepared calcium solutions (water- and ethanol-based) were added dropwise into the phosphate solutions (water- and ethanol-based) under stirring, with the pH adjusted to 11 using ammonium hydroxide. The solutions were then aged for 24 hours at room temperature and exposed to thermal treatment at 105°C in an air oven until white-dried gels were obtained. The dried powders were ground in a mortar and calcined for 1, 2, and 4 hours at 950°C, with a heating rate of 5°C/min.

XRD characterization revealed that, despite the calcination time, the samples obtained by the ethanol-based synthesis presented calcium oxide and calcium hydroxide as impurities, while with the water-based synthesis, the only secondary phase was calcium oxide, with the 1-hour calcination yielding the best result.

The SEM analysis showed that the powders obtained from both solvents were non-uniform, and the arrangement of cluster-like particles formed nano-sized spherical aggregates for all calcination times. However, it can be said that, to some degree, homogeneous particles with a narrow size range and lower degree of aggregation occurred in the water-based samples.

Farazin et al. [90] synthesized HAp to develop a chitosan/hydroxyapatite composite scaffold for use as a bone replacement tissue. The authors used diammonium hydrogen phosphate and calcium nitrate tetrahydrate as precursors, with a Ca/P molar ratio of 1.67. Diammonium hydrogen phosphate was dissolved in deionized water, and calcium nitrate tetrahydrate was dissolved in a mixture of ethanol and deionized water in a 1:12 ratio. Ammonia solution was used to maintain the pH between 10 and 12. The first solution was added to the second one, with this step being carried out in a low-temperature water bath (38–42°C) for 4 hours under magnetic stirring. The resulting material was filtered and washed several times with deionized water.

The drying process comprised three steps: 14 hours at 40°C, 2 hours at 60°C, and finally 18 hours at 80°C, followed by heat treatment at 700°C for 1 hour, with a heating rate of 10°C/min.

XRD analysis showed that, besides HAp, beta-tricalcium phosphate and calcium phosphate were formed as secondary phases. On the other hand, FESEM micrographs revealed that a nanostructured powder was obtained, with an average size of 48 nm (**Table 2**).

6. Case study

This case study explores the synthesis and characterization of HAp. For the synthesis, the sol-gel route proposed by Figueroa et al. [87] was used. However, to enable upscaling of the process, the mass of the precursors was doubled along with the volume of the solvent. The obtained powder was thermally analyzed and heat-treated at four different temperatures. Structural characterization of the four samples was performed using X-ray diffraction, while the morphology and electrical behavior of the most promising sample were analyzed using scanning electron microscopy and impedance spectroscopy, respectively.

6.1 Materials and methods

6.1.1 Materials

For the hydroxyapatite synthesis, the starting materials were calcium nitrate tetrahydrate ($Ca(NO_3)_2{\cdot}4H_2O$, EssentQ®, Scharlab), diphosphorus pentoxide (P_2O_5, EssentQ®, Scharlab), and ethanol (C_2H_5OH, 96% v/v, Aga) as the solvent.

6.1.2 Synthesis of HAp

As mentioned, the synthesis of HAp was performed using the sol-gel method, following a modified approach originally presented by Figueroa et al. [87]. The first solution was prepared by dissolving 8.52 g of P_2O_5 in 175.0 ml of ethanol, and the second solution was prepared by dissolving 47.20 g of $Ca(NO_3)_2{\bullet}4H_2O$ in 584.0 ml of ethanol in order to achieve a stoichiometric Ca:P ratio of 1.67.

Afterward, the first solution was added to the second solution in a dropwise fashion. The final solution was stirred for an hour at room temperature and then dried at 80°C for 36 hours to evaporate the solvent.

To evaluate the temperatures at which crystalline phases form, the obtained powder was subjected to thermal characterization using differential scanning calorimetry and thermogravimetric analysis. Based on these results, and after grinding, the powder was heat-treated at 500, 600, and 700°C for 2 hours, with a heating rate of 5°C/min.

Figure 9 shows the flowchart of the steps involved in the synthesis of the HAp powders.

6.1.3 Characterization techniques

The thermogravimetric analysis, TGA, was performed on a Netzsch TG 209 F1 Libra in a nitrogen atmosphere, with a heating rate of 5°C/min, ranging from 30 to 1095°C. The differential scanning calorimetry, DSC, was carried out between 25 and 500°C with a Netzsch DSC 204 F1 Phoenix, also in a nitrogen atmosphere, at the same heating rate of 5°C/min.

X-ray Diffraction, XRD, analyses were carried out using a PANalytical X'Pert PRO diffractometer with Cu Kα (λ = 1.54060 Å) radiation, operating at 45 kV and 40 mA. The tests utilized parallel beam geometry, with a step size of 0.04° and an exposure time of 3 s per step.

The morphology analysis was conducted by Scanning Electron Microscopy, SEM, utilizing a TESCAN Vega 3 microscope. All measurements were taken at a voltage of 15 kV.

The impedance spectroscopy analysis was made on the sample having a disk shape with a diameter of 13.10 mm and a thickness of approximately 1.36 mm. Silver electrodes were prepared by applying a thin layer of silver conductive ink onto the opposite sides of the sample.

The measurements were performed using an Agilent 4294A precision impedance analyzer in C_p–R_p configuration (capacitance in parallel with resistance), with an applied voltage of 0.5 V, at different temperatures over the frequency range of 40 Hz to 1 MHz. During these measurements, the sample was maintained in a helium atmosphere to improve the heat transfer and eliminate moisture.

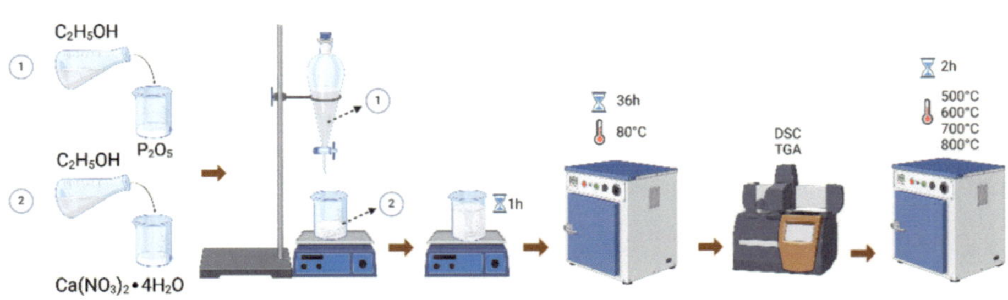

Figure 9.
Schematic representation of the HAp synthesis process [Created in BioRender. https://BioRender.com/x36v638].

6.2 Results

6.2.1 Thermal, structural, and morphological characterization

Figure 10(a) shows the TGA curve of the HAp powder at a heating rate of 5°C/min, revealing a total weight loss of 45.94%, with the most significant mass loss occurring below 500°C.

In the first stage, between 33 and 79°C, the weight loss can be attributed to the evaporation of physically adsorbed ethanol and water, which corresponds to the first endothermic peak in the DSC curve, **Figure 10(b)**, centered at 52°C [91–95].

The second stage, in the range of 79 to 360°C, exhibits weight loss due to the release of adsorbed and lattice water, as well as the partial decomposition of organic matter and crystallization of HAp. The release of the adsorbed and lattice water corresponds to the endothermic peaks in the DSC curve, centered at approximately 118 and 140°C, while the decomposition of organic matter and crystallization of HAp can be assigned exothermic peaks centered at 310 and 341°C, respectively [91, 94, 96–98].

According to Tõnsuaadu et al. [99], the lattice water is irreversibly lost between 200 and 400°C. Such temperature variations can be ascribed to differences in the samples and the experimental synthesis procedures [93, 100].

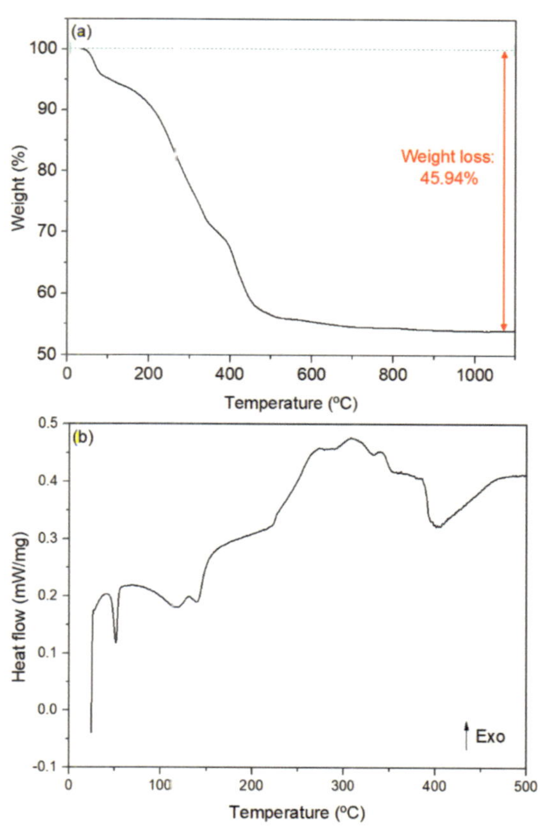

Figure 10.
(a) TGA and (b) DSC curves of the obtained powder.

The third stage, occurring between 360 and 500°C, likely represents the loss of water due to the decomposition of chemically bonded water, as well as the decomposition of the organic matrix and carbonates. In the DSC curve, these events correspond to the broad endothermic peak centered at 425°C [91, 95, 101]. Above 500°C, the weight loss becomes minimal.

Based on this analysis, heat treatments at 500, 600, 700, and 800°C were selected to observe the progression from crystal growth and decomposition at lower temperatures to the stability and crystallization expected at higher temperatures.

The crystalline phases were identified by XRD, as shown in **Figure 11(a)**. In addition to the patterns of the sintered powders, the reference pattern for HAp (ICDD 01-074-0565 [102]) is also displayed.

The formation of hexagonal HAp is observed even at the lowest heat treatment temperature of 500°C. However, along with HAp, three secondary phases—$Ca_2P_2O_7$, $Ca_3(PO_4)_2$, and Ca_3N_2—were identified. These phases were also present in the samples heat-treated at 600 and 700°C, with the heat treatment at 800°C resulting in the decomposition of both HAp and Ca_3N_2.

The chemical composition of the sintered samples was estimated using Rietveld refinement, performed with the software Profex [103], with the goodness of fit ranging from 1.39 to 1.81. The weight percentage of each phase is presented in **Figure 11(b)**, showing that as temperature increases, the HAp content decreases.

These results indicate that when using the approach proposed by Figueroa et al. [87] as a starting point, the upscaling of the process cannot be governed solely by the concentration of the precursor solutions. Using the same concentrations reported by Figueroa et al., but doubling the mass of calcium nitrate tetrahydrate and diphosphorus pentoxide and the corresponding ethanol volumes, the efficiency of the process remains lower than that achieved by the mentioned authors, who reported HAp as the main component with only a small trace of calcium phosphate.

Based on the obtained results, the sample heat-treated at 500°C demonstrated the most favorable properties, making it the most suitable candidate for further characterization.

Figure 12 presents the SEM micrographs of this sample captured at two different magnifications, providing insights into its surface morphology. Signs of agglomeration are visible, along with particles of varying shapes and highly diverse sizes, ranging from the nanometric to micrometric scale.

6.2.2 Impedance spectroscopy measurements

Impedance spectroscopy is a reliable technique for quantitatively interpreting the electrical properties of materials and studying the processes occurring within them. It enables a direct correlation between the response of a real system and an idealized model circuit composed of discrete electrical components, allowing the separation of grain and grain boundary contributions to the transport properties of materials [104, 105].

Since $Z^*(\omega)$ is a complex quantity, it can be represented in polar coordinates or Cartesian coordinates. This representation is given by [106]:

$$Z^* = Z' - jZ'' \tag{2}$$

where Z' and Z'' are the real and imaginary parts of the complex impedance, respectively, and can be calculated using the equations [105]:

$$Z' = \frac{R_P}{1+(\omega R_P C_P)^2} \tag{3}$$

$$Z'' = \frac{\omega R_P{}^2 C_P}{1+(\omega R_P C_P)^2} \tag{4}$$

where C_p and R_p are the measured capacitance and resistance, respectively, and ω is the angular frequency.

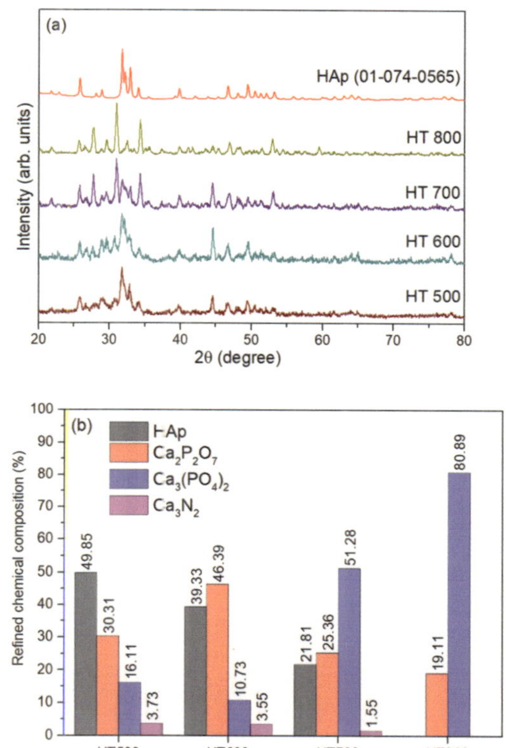

Figure 11.
(a) XRD patterns and (b) chemical composition, in weight percentage, of the heat-treated powders.

Figure 12.
SEM micrographs of the samples heat-treated at 500°C.

Figure 13 shows the imaginary part of the impedance, Z'', as a function of frequency at different temperatures.

Analyzing the variation of Z'' with the frequency, one can infer the appearance of peaks at a particular frequency, which became evident for temperatures above 300 K, and the convergence of the spectra at higher frequencies.

This behavior suggests the presence of a thermally activated relaxation mechanism in the system in the frequency and temperature ranges studied [107].

In addition to complex impedance, the modulus formalism can also be applied to study the dielectric behavior of a ceramic material, with several authors emphasizing the advantages of the combined use of both formalisms, specifically because the Z'' plots highlight the phenomenon of the largest resistance, whereas M'' highlights those of the smallest capacitance [108, 109].

The dielectric modulus, M^*, is defined as a quantity directly proportional to the complex impedance [108, 110]:

$$M^* = M' + jM'' = j\omega C_0 Z^* \tag{5}$$

where M' and M'' are the real and imaginary parts of the complex modulus, respectively, and can be obtained from [111]:

$$M' = -\omega C_0 Z'' \tag{6}$$

$$M'' = \omega C_0 Z' \tag{7}$$

C_0 is the capacitance of the empty (vacuum) measurement cell and is given by [112]:

$$C_0 = \varepsilon_0 \frac{A}{d} \tag{8}$$

where ε_0 is the vacuum permittivity (8.8542×10^{-12} F/m), d is the sample thickness, and A is the electrode area.

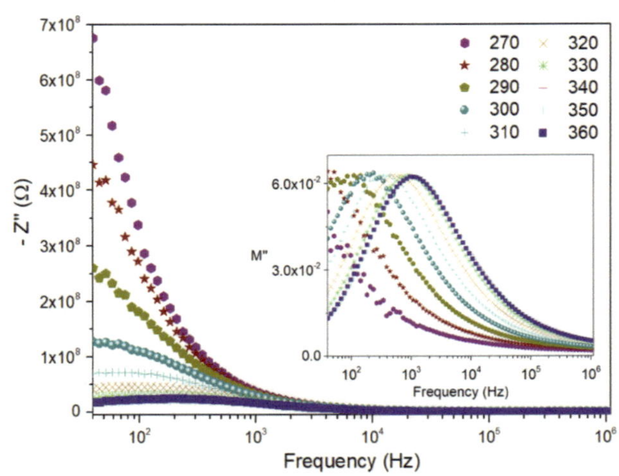

Figure 13.
The imaginary part of the impedance, Z″, as a function of frequency, for temperatures between 270 and 360 K, in steps of 10 K, for the sample heat-treated at 500°C (inset: Imaginary part of the modulus, M″, as a function of frequency, in the same temperature range).

In the inset of **Figure 13**, the imaginary part of the modulus is depicted as a function of frequency at the same temperatures. The frequency dependence of M'' clearly exhibits a maximum for each temperature, once again above 300 K. It is observed that the position of the peak shifts toward higher frequencies as the temperature increases, confirming the relaxation mechanism previously mentioned.

Figure 14 shows the plots of the imaginary part of impedance and modulus versus frequency at 300 K. It can be seen from the figure that Z''_{max} and M''_{max} do not occur at the same frequency, indicating non-Debye-type relaxation [113].

Figure 15 shows the complex impedance plots, Z'' as a function of Z'. In the temperature range previously explored, the Nyquist plots consist of semicircular arcs. For low temperatures, the large radius indicates the high-insulating nature of the material. With further increasing temperature, the radius of the distorted semicircular arc decreases, indicating that the conductivity increases with increasing temperature. For higher temperatures, the curves bend toward the real Z'-axis to form distorted semicircular arcs that support the non-Debye-type relaxation phenomenon.

The distorted nature of the semicircular arcs, as highlighted in the inset for 300 K, may imply the existence of multi relaxation processes in the sample, which were not revealed by the previous analyses. Also, in the inset of **Figure 15**, the Bode-phase diagram allows us to infer the existence of one additional relaxation process in the low-frequency region, since a step-like change is starting to appear [114]. To make a clearer understanding of the multirelaxation behavior in the sample, a simulation of complex impedance data was performed with equivalent electric circuits.

Often, a single data set can be reasonably fitted by multiple equivalent circuits, making the choice dependent on both simplicity and consistency with the known physical and chemical processes occurring in the system [115].

In this study, the impedance data was modeled employing an equivalent circuit, containing a series combination of two parallel (R//CPE) elements (as also depicted in the inset of **Figure 15**), having the model circuit parameters been obtained using the EIS Spectrum Analyzer fitting software [116]. The simulated response of the equivalent circuit for 300 K is also shown in the inset of **Figure 15**.

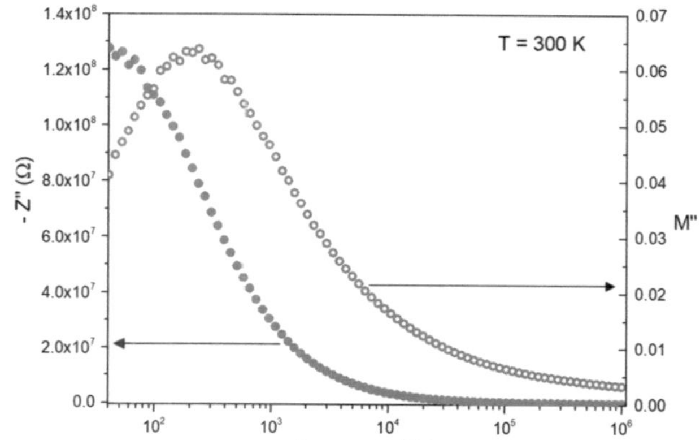

Figure 14.
The imaginary part of the impedance and modulus, as a function of frequency, at T = 300 K.

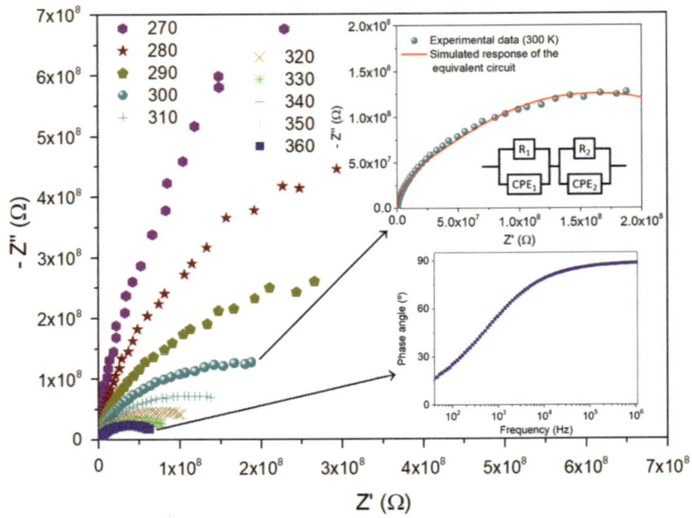

Figure 15.
Nyquist plots for temperatures between 270 and 360 K, in steps of 10 K (inset: Nyquist plot and its fit for T = 300 K and the equivalent circuit model considered; Bode-phase diagram for T = 300 K).

The impedance of a constant phase element is given by:

$$Z = Af^{-n}\left(\cos\left(\frac{n\pi}{2}\right) - j\sin\left(\frac{n\pi}{2}\right)\right) \tag{9}$$

where A is the impedance observed for the angular velocity of 1 rad/s, f is the frequency, and n is an exponent equal to 1 for the case of a pure capacitor. The CPE was used to replace capacitance to account for non-ideal Debye behavior [117].

The semicircular arc at high frequencies arises mainly from the grain properties, while the appearance of a second semicircle at a lower frequency can be attributed to the grain boundary properties [118].

Table 3 summarizes the estimated resistance and capacitance values by fitting the experimental data using the mentioned software.

6.3 Conclusions

In summary, this study demonstrates that the modification of a pre-existing synthesis method highlights the complexity of the process, emphasizing the need for careful optimization of precursor concentrations to achieve the desired phase purity. The highest content of HAp was obtained with a heat treatment at 500°C, with three secondary phases, $Ca_2P_2O_7$, $Ca_3(PO_4)_2$, and Ca_3N_2, being identified. The morphology of the sample was characterized by particles of varying shapes and sizes, with signs of agglomeration. Regarding the electrical properties, the impedance data were modeled using an equivalent circuit, which consisted of a series combination of two parallel

T (K)	R_1 (MΩ)	R_2 (MΩ)	Q_1 (pS)	n_1 (a.u.)	Q_2 (pS)	n_2 (a.u.)
300	49.61	245.63	11.13	0.966	13.78	0.977

Table 3.
Fitted results for the grain and grain boundary resistance, R_1 and R_2, and corresponding admittances, Q_1 and Q_2, of the CPE elements, at 300 K.

(R//CPE) elements, with the grain and grain boundary resistance presenting values of 46.61 and 245.63 MΩ, respectively, at 300 K.

7. Final remarks and outlooks about HAp synthesis methods

Over the years, substantial progress has elevated HAp from a basic biocompatible material to a sophisticated functional substance with diverse applications.

The synthesis of HAp is a multifaceted field that encompasses a wide range of methods, each with its advantages and limitations. Dry methods, such as solid-state and mechanochemical synthesis, offer scalability and simplicity but often yield products with lower purity and irregular morphology. Wet methods, including chemical precipitation, hydrolysis, and sol-gel synthesis, enable precise control over particle size and composition, making them suitable for high-performance applications. High-temperature methods and hybrid techniques further expand the toolkit for HA synthesis, providing avenues for producing materials with tailored properties.

Although numerous methods have been explored for HAp synthesis, only a few are assumed viable when considering both performance and economic factors. The primary challenges associated with most techniques include the reliance on various precursors, solvents, or surfactants, along with issues such as broad particle size distribution, complex and costly procedures, significant agglomeration, and the presence of phase impurities.

On the other hand, biogenic synthesis represents a promising direction for sustainable HAp production, aligning with the global emphasis on resource conservation and waste valorization. Biogenic synthesis involves deriving HAp from natural sources such as animal bones, eggshells, and plant materials. These methods capitalize on the natural composition of biogenic materials, reducing the need for synthetic precursors. Techniques such as thermal calcination and chemical precipitation are employed to extract and process HAp from these sources. Biogenic HAp often contains trace elements that enhance its bioactivity. Integrating advanced technologies like microwave-assisted synthesis and additive manufacturing has the potential to revolutionize the production and application of HAp in biomedical engineering.

Acknowledgements

This research is sponsored by national funds through FCT—Fundação para a Ciência e Tecnologia, under projects UIDB/00285—Centre for Mechanical Engineering, Materials and Processes and LA/P/0112/2020, and by i3N (LA/P/0037/2020, UID-B/50025/2020, and UID-P/50025/2020) financed by the COMPETE 2020 Program and national funds through FCT/MEC and FEDER under the PT2020 Partnership Agreement.

Conflict of interest

The authors declare no conflict of interest.

Author details

Susana Devesa[1*], Zohra Benzarti[1,2], Sílvia Soreto[3] and Sandra Carvalho[1]

1 CEMMPRE, ARISE, Department of Mechanical Engineering, University of Coimbra, Coimbra, Portugal

2 Laboratory of Multifunctional Materials and Applications (LaMMA), Department of Physics, Faculty of Sciences of Sfax, University of Sfax, Sfax, Tunisia

3 i3N and Department of Physics, University of Aveiro, Campus Universitário de Santiago, Aveiro, Portugal

*Address all correspondence to: susana_devesa@hotmail.com

References

[1] Haider A, Haider S, Han SS, Kang IK. Recent advances in the synthesis, functionalization and biomedical applications of hydroxyapatite: A review. RSC Advances. 2017;7(13):7442-7458

[2] Koutsopoulos S. Synthesis and characterization of hydroxyapatite crystals: A review study on the analytical methods. Journal of Biomedical Materials Research. 2002;**62**(4):600-612

[3] Rocha RLP, Honorio LMC, RDS DB, Trigueiro P, Duarte TM, Fonseca MG, et al. Light-activated hydroxyapatite photocatalysts: New environmentally-friendly materials to mitigate pollutants. Minerals. 2022;**12**(5):525

[4] Dorozhkin SVA. Detailed history of calcium orthophosphates from 1770s till 1950. Materials Science and Engineering: C. 2013;**33**(6):3085-3110

[5] Shepperd J. The early biological history of calcium phosphates. In: Fifteen Years of Clinical Experience with Hydroxyapatite Coatings in Joint Arthroplasty. Paris: Springer; 2004. pp. 3-8

[6] Dorozhkin SV. Calcium orthophosphates. Biomatter. 2011;**1**(2):121-164

[7] Dorozhkin SV. Calcium orthophosphates in nature, biology and medicine. Materials. 2009;**2**(2):399-498

[8] Dorozhkin SV. Calcium orthophosphates. Journal of Materials Science. 2007;**42**(4):1061-1095

[9] Ginebra MP. Calcium phosphate bone cements. In: Orthopaedic Bone Cements. Cambridge, United Kingdom: Elsevier; 2008. pp. 206-230

[10] Bystrov VS, Coutinho J, Bystrova AV, Dekhtyar YD, Pullar RC, Poronin A, et al. Computational study of hydroxyapatite structures, properties and defects. Journal of Physics D. 2015;**48**(19):195302

[11] Elliott JC, Mackie PE. Young RA monoclinic hydroxyapatite. Science (1979). 1973;**180**(4090):1055-1057

[12] Slepko A, Demkov AA. Hydroxyapatite: Vibrational spectra and monoclinic to hexagonal phase transition. Journal of Applied Physics. 2015;**117**(7):074701

[13] Ma G, Liu XY. Hydroxyapatite: Hexagonal or monoclinic? Crystal Growth and Design. 2009;**9**(7):2991-2994

[14] Mathew M, Takagi S. Structures of biological minerals in dental research. Journal of Research of the National Institute of Standards and Technology. 2001;**106**(6):1035

[15] Khan AS, Chaudhry AA. Analytical tools for substituted hydroxyapatite. In: Raza M, Zahid S, Asif A, editors. Handbook of Ionic Substituted Hydroxyapatites. Cambridge, United Kingdom: Woodhead Publishing; 2019

[16] El Feki H, Savariault JM, Ben SA. Structure refinements by the Rietveld method of partially substituted hydroxyapatite: $Ca_9Na_{0.5}(PO_4)_{4.5}(CO_3)_{1.5}(OH)_2$. Journal of Alloys and Compounds. 1999;**287**(1-2):114-120

[17] Rangavittal N, Landa-Cánovas AR, González-Calbet JM, Vallet-Regí M. Structural study and stability of hydroxyapatite and ?-tricalcium phosphate: Two important bioceramics.

Journal of Biomedical Materials Research. 2000;**51**(4):660-668

[18] Zilm ME, Chen L, Sharma V, McDannald A, Jain M, Ramprasad R, et al. Hydroxyapatite substituted by transition metals: Experiment and theory. Physical Chemistry Chemical Physics. 2016;**18**(24):16457-16465

[19] Bystrov VS. Piezoelectricity and pyroelectricity in hydroxyapatite. Ferroelectrics. 2019;**541**(1):25-29

[20] Boanini E, Gazzano M, Bigi A. Ionic substitutions in calcium phosphates synthesized at low temperature. Acta Biomaterialia. 2010;**6**(6):1882-1894

[21] NAS M P'a, Koshy P, Abdullah HZ, Idris MI, Lee TC. Syntheses of hydroxyapatite from natural sources. Heliyon. 2019;**5**(5):e01588

[22] Rial R, Tahoces PG, Hassan N, Cordero ML, Liu Z, Ruso JM. Noble microfluidic system for bioceramic nanoparticles engineering. Materials Science and Engineering: C. 2019;**102**:221-227

[23] Mondal S, Dorozhkin SV, Pal U. Recent progress on fabrication and drug delivery applications of nanostructured hydroxyapatite. WIREs Nanomedicine and Nanobiotechnology. 2018;**10**(4):e1504

[24] Zhou H, Lee J. Nanoscale hydroxyapatite particles for bone tissue engineering. Acta Biomaterialia. 2011;**7**(7):2769-2781

[25] Hoveidaei AH, Sadat-Shojai M, Mosalamiaghili S, Salarikia SR, Roghani-shahraki H, Ghaderpanah R, et al. Nano-hydroxyapatite structures for bone regenerative medicine: Cell-material interaction. Bone. 2024;**179**:116956

[26] Lara-Ochoa S, Ortega-Lara W, Guerrero-Beltrán CE. Hydroxyapatite nanoparticles in drug delivery: Physicochemistry and applications. Pharmaceutics. 2021;**13**(10):1642

[27] Hartati YW, Irkham I, Zulqaidah S, Syafira RS, Kurnia I, Noviyanti AR, et al. Recent advances in hydroxyapatite-based electrochemical biosensors: Applications and future perspectives. Sensing and Bio-Sensing Research. 2022;**38**:100542

[28] Huang J, Lin YW, Fu XW, Best SM, Brooks RA, Rushton N, et al. Development of nano-sized hydroxyapatite reinforced composites for tissue engineering scaffolds. Journal of Materials Science. Materials in Medicine. 2007;**18**(11):2151-2157

[29] Li S, Li Y, Shen W, Bai Y, Kong L. Hydroxyapatite-based catalysis in environmental decontamination. Journal of Cleaner Production. 2022;**380**:134961

[30] Townsend L, Williams RL, Anuforom O, Berwick MR, Halstead F, Hughes E, et al. Antimicrobial peptide coatings for hydroxyapatite: Electrostatic and covalent attachment of antimicrobial peptides to surfaces. The Journal of the Royal Society Interface. 2017;**14**(126):20160657

[31] Predoi D, Iconaru SL, Predoi MV. Fabrication of silver- and zinc-doped hydroxyapatite coatings for enhancing antimicrobial effect. Coatings. 2020;**10**(9):905

[32] Wang Y, Liang D, Liu F, Zhang W, Di X, Wang C. A polyethylene glycol/hydroxyapatite composite phase change material for thermal energy storage. Applied Thermal Engineering. 2017;**113**:1475-1482

[33] Ramesan MT, Sankar S, Kalladi AJ, Labeeba Abdulla AC, Bahuleyan BK.

Hydroxyapatite nanoparticles reinforced polyvinyl alcohol/chitosan blend for optical and energy storage applications. Polymer Engineering and Science. 2024;**64**(3):1378-1390

[34] Khtaoui L, Laghrouche M, Fernane F, Chaouchi A. High-sensitivity humidity sensor based on natural hydroxyapatite. Journal of Materials Science: Materials in Electronics. 2021;**32**(7):8668-8686

[35] Mondal S, Hoang G, Manivasagan P, Kim H, Oh J. Nanostructured hollow hydroxyapatite fabrication by carbon templating for enhanced drug delivery and biomedical applications. Ceramics International. 2019;**45**(14):17081-17093

[36] Machado TR, Leite IS, Inada NM, Li MS, da Silva JS, Andrés J, et al. Designing biocompatible and multicolor fluorescent hydroxyapatite nanoparticles for cell-imaging applications. Materials Today Chemistry. 2019;**14**:100211

[37] Kargozar S, Mollazadeh S, Kermani F, Webster TJ, Nazarnezhad S, Hamzehlou S, et al. Hydroxyapatite nanoparticles for improved cancer. Journal of Functional Biomaterials. 2022;**13**(3):100

[38] Damiri F, Fatimi A, Magdalena Musuc A, Paiva Santos AC, Paszkiewicz S, Igwe Idumah C, et al. Nano-hydroxyapatite (nHAp) scaffolds for bone regeneration: Preparation, characterization and biological applications. Journal of Drug Delivery Science and Technology. 2024;**95**:105601

[39] Sadat-Shojai M, Khorasani MT, Dinpanah-Khoshdargi E, Jamshidi A. Synthesis methods for nanosized hydroxyapatite with diverse structures. Acta Biomaterialia. 2013;**9**(8):7591-7621

[40] Laonapakul T, Sutthi R, Chaikool P, Talangkun S, Boonma A, Chindaprasirt P. Calcium phosphate powders synthesized from $CaCO_3$ and CaO of natural origin using mechanical activation in different media combined with solid-state interaction. Materials Science and Engineering: C. 2021;**118**:111333

[41] Bulina NV, Vinokurova OB, Eremina NV, Prosanov IY, Khusnutdinov VR, Chaikina MV. Features of solid-phase mechanochemical synthesis of hydroxyapatite doped by copper and zinc ions. Journal of Solid State Chemistry. 2021;**296**:121973

[42] Gupta SK, Mao Y. Recent developments on molten salt synthesis of inorganic nanomaterials: A review. The Journal of Physical Chemistry C. 2021;**125**(12):6508-6533

[43] Gupta SK, Mao Y. A review on molten salt synthesis of metal oxide nanomaterials: Status, opportunity, and challenge. Progress in Materials Science. 2021;**117**:100734

[44] Li S, Liu J, Wang ZL, Wei D. Mechano-driven chemical reactions. Green Energy and Environment. 2024

[45] Fihri A, Len C, Varma RS, Solhy A. Hydroxyapatite: A review of syntheses, structure and applications in heterogeneous catalysis. Coordination Chemistry Reviews. 2017;**347**:48-76

[46] Gomez-Vazquez OM, Correa-Piña BA, Zubieta-Otero LF, Castillo-Paz AM, Londoño-Restrepo SM, Rodriguez-García ME. Synthesis and characterization of bioinspired nano-hydroxyapatite by wet chemical precipitation. Ceramics International. 2021;**47**(23):32775-32785

[47] Lee IH, Lee JA, Lee JH, Heo YW, Kim JJ. Effects of pH and reaction temperature on hydroxyapatite powders synthesized by precipitation.

Journal of the Korean Ceramic Society. 2020;**57**(1):56-64

[48] Nzereogu PU, Omah AD, Ezema FI, Iwuoha EI, Nwanya AC. Silica extraction from rice husk: Comprehensive review and applications. Hybrid Advances. 2023;**4**:100111

[49] Gentile P, Wilcock C, Miller C, Moorehead R, Hatton P. Process optimisation to control the physico-chemical characteristics of biomimetic nanoscale hydroxyapatites prepared using wet chemical precipitation. Materials. 2015;**8**(5):2297-2310

[50] Ahmed MK, Mansour SF, Al-Wafi R, Afifi M, Uskoković V. Gold as a dopant in selenium-containing carbonated hydroxyapatite fillers of nanofibrous ε-polycaprolactone scaffolds for tissue engineering. International Journal of Pharmaceutics. 2020;**577**:118950

[51] Zhang L, Lu T, He F, Zhang W, Yuan X, Wang X, et al. Physicochemical and cytological properties of poorly crystalline calcium-deficient hydroxyapatite with different Ca/P ratios. Ceramics International. 2022;**48**(17):24765-24776

[52] Suchanek WL, Riman RE. Hydrothermal synthesis of advanced ceramic powders. Advances in Science and Technology. 2006;**45**:184-193

[53] Sadat-Shojai M. Preparation of hydroxyapatite nanoparticles: Comparison between hydrothermal and solvo-treatment processes and colloidal stability of produced nanoparticles in a dilute experimental dental adhesive. Journal of the Iranian Chemical Society. 2009;**6**(2):386-392

[54] Vijayalakshmi U, Prabakaran K, Rajeswari S. Preparation and characterization of sol–gel hydroxyapatite and its electrochemical evaluation for biomedical applications. Journal of Biomedical Materials Research Part A. 2008;**87A**(3):739-749

[55] Lett JA, Sundareswari M, Ravichandran K, Latha MB, Sagadevan S, MohdR BJ. Tailoring the morphological features of sol–gel synthesized mesoporous hydroxyapatite using fatty acids as an organic modifier. RSC Advances. 2019;**9**(11):6228-6240

[56] Phatai P, Futalan CM, Kamonwannasit S, Khemthong P. Structural characterization and antibacterial activity of hydroxyapatite synthesized via sol-gel method using glutinous rice as a template. Journal of Sol-Gel Science and Technology. 2019;**89**(3):764-775

[57] Murray MGS, Wang J, Ponton CB, Marquis PM. An improvement in processing of hydroxyapatite ceramics. Journal of Materials Science. 1995;**30**(12):3061-3074

[58] Zhou WY, Wang M, Cheung WL, Guo BC, Jia DM. Synthesis of carbonated hydroxyapatite nanospheres through nanoemulsion. Journal of Materials Science: Materials in Medicine. 2008;**19**(1):103-110

[59] Pradeesh TS, Sunny MC, Varma HK, Ramesh P. Preparation of microstructured hydroxyapatite microspheres using oil in water emulsions. Bulletin of Materials Science. 2005;**28**(5):383-390

[60] Wei K, Wang Y, Lai C, Ning C, Wu D, Wu G, et al. Synthesis and characterization of hydroxyapatite nanobelts and nanoparticles. Materials Letters. 2005;**59**(2-3):220-225

[61] Jevtić M, Mitrić M, Škapin S, Jančar B, Ignjatović N, Uskoković D.

Crystal structure of hydroxyapatite nanorods synthesized by sonochemical homogeneous precipitation. Crystal Growth and Design. 2008;**8**(7):2217-2222

[62] Han Y, Wang X, Li S. A simple route to prepare stable hydroxyapatite nanoparticles suspension. Journal of Nanoparticle Research. 2009;**11**(5):1235-1240

[63] Kim W, Saito F. Sonochemical synthesis of hydroxyapatite from H_3PO_4 solution with $Ca(OH)_2$. Ultrasonics Sonochemistry. 2001;**8**(2):85-88

[64] Rouhani P, Taghavinia N, Rouhani S. Rapid growth of hydroxyapatite nanoparticles using ultrasonic irradiation. Ultrasonics Sonochemistry. 2010;**17**(5):853-856

[65] Kuznetsov AV, Fomin AS, Veresov AG, Putlyaev VI, Fadeeva IV, Barinov SM. Hydroxyapatite of platelet morphology synthesized by ultrasonic precipitation from solution. Russian Journal of Inorganic Chemistry. 2008;**53**(1):1-5

[66] Kavitha M, Subramanian R, Vinoth KS, Narayanan R, Venkatesh G, Esakkiraja N. Optimization of process parameters for solution combustion synthesis of strontium substituted hydroxyapatite nanocrystals using Design of Experiments approach. Powder Technology. 2015;**271**:167-181

[67] Aizawa M, Hanazawa T, Itatani K, Howell FS, Kishioka A. Characterization of hydroxyapatite powders prepared by ultrasonic spray-pyrolysis technique. Journal of Materials Science. 1999;**34**:2865-2873

[68] Alam K, Hossain K, Kawsar BNM, Ahmed S. Synthesis of nano-hydroxyapatite using emulsion, pyrolysis, combustion, and sonochemical methods and biogenic sources: A review. RSC Advances. 2024;**14**:3548-3559

[69] Wakiya N, Yamasaki M, Adachi T, Inukai A, Sakamoto N, Fu D, et al. Preparation of hydroxyapatite–ferrite composite particles by ultrasonic spray pyrolysis. Materials Science and Engineering: B. 2010;**173**:195-198

[70] Sasikumar S, Vijayaraghavan R. Solution combustion synthesis of bioceramic calcium phosphates by single and mixed fuels—A comparative study. Ceramics International. 2008;**34**(6):1373-1379

[71] Canillas M, Rivero R, García-Carrodeguas R, Barba F, Rodríguez MA. Processing of hydroxyapatite obtained by combustion synthesis. Boletín de la Sociedad Española de Cerámica y Vidrio. 2017;**56**(5):237-242

[72] Kavitha M, Subramanian R, Narayanan R, Udhayabanu V. Solution combustion synthesis and characterization of strontium substituted hydroxyapatite nanocrystals. Powder Technology. 2014;**253**:129-137

[73] Narayanan R, Singh V, Kwon TY, Kim KH. Combustion synthesis of hydroxyapatite and hydroxyapatite (silver) powders. Key Engineering Materials. 2008;**396-398**:411-419

[74] Putro JN, Handoyo N, Kristiani V, Soenjaya SA, Ki OL, Soetaredjo FE, et al. Pomacea sp shell to hydroxyapatite using the ultrasound–microwave method (U–M). Ceramics International. 2014;**40**(7):11453-11456

[75] Sri Ramakrishnan L, Unnikrishnan PS, Sabu CK, Krishnan AG, Nair MB. Effect of wheat gluten on improved thermal cross-linking and osteogenesis of hydroxyapatite-gelatin composite scaffolds. International

Journal of Biological Macromolecules. 2021;**183**:1200-1209

[76] Nazeer MA, Yilgor E, Yagci MB, Unal U, Yilgor I. Effect of reaction solvent on hydroxyapatite synthesis in sol–gel process. Royal Society Open Science. 2017;**4**(12):171098

[77] Etinosa PO, Osuchukwu OA, Anisiji EO, Lawal MY, Mohammed SA, Ibitoye OI, et al. In-depth review of synthesis of hydroxyapatite biomaterials from natural resources and chemical regents for biomedical applications. Arabian Journal of Chemistry. 2024;**17**(12):106010

[78] Nayak AK. Hydroxyapatite synthesis methodologies: An overview. International Journal of Chem Tech Research. 2010;**2**(2):903-907

[79] Anbalagan Balamurugan JM, Faure J, Hicham Benhayoune L, Sockallingum G, Banchet V, Bouthors S, et al. Synthesis and structural analysis of sol-gel derived stoichiometric monophasic hydroxyapatit. Ceramics-Silikaty. 2006;**50**(1):27-31

[80] Li R, Clark AE, Hench LL. An investigation of bioactive glass powders by sol-gel processing. Journal of Applied Biomaterials. 1991;**2**(4):231-239

[81] Kuriakose TA, Kalkura SN, Palanichamy M, Arivuoli D, Dierks K, Bocelli G, et al. Synthesis of stoichiometric nano crystalline hydroxyapatite by ethanol-based sol–gel technique at low temperature. Journal of Crystal Growth. 2004;**263**(1-4):517-523

[82] Liu DM, Troczynski T, Tseng WJ. Water-based sol–gel synthesis of hydroxyapatite: Process development. Biomaterials. 2001;**22**(13):1721-1730

[83] Livage J, Barboux P, Vandenborre MT, Schmutz C, Taulelle F. Sol-gel synthesis of phosphates. Journal of Non-Crystalline Solids. 1992;**147-148**:18-23

[84] Bogdanoviciene I, Beganskiene A, Tõnsuaadu K, Glaser J, Meyer HJ, Kareiva A. Calcium hydroxyapatite, $Ca_{10}(PO_4)_6(OH)_2$ ceramics prepared by aqueous sol–gel processing. Materials Research Bulletin. 2006;**41**(9):1754-1762

[85] Padmanabhan SK, Balakrishnan A, Chu MC, Lee YJ, Kim TN, Cho SJ. Sol–gel synthesis and characterization of hydroxyapatite nanorods. Particuology. 2009;**7**(6):466-470

[86] Chen J, Wang Y, Chen X, Ren L, Lai C, He W, et al. A simple sol-gel technique for synthesis of nanostructured hydroxyapatite, tricalcium phosphate and biphasic powders. Materials Letters. 2011;**65**(12):1923-1926

[87] Figueroa IA, Novelo-Peralta O, Flores-Morales C, González-Tenorio R, Piña-Barba MC. Synthesis and characterization of biocompatible-nanohydroxyapatite crystals obtained by a modified sol-gel processing. Biomatter. 2012;**2**(2):71-76

[88] Bakan F, Laçin O, Sarac H. A novel low temperature sol–gel synthesis process for thermally stable nano crystalline hydroxyapatite. Powder Technology. 2013;**233**:295-302

[89] Türk S, Altınsoy İ, Çelebi Efe G, Ipek M, Özacar M, Bindal C. Effect of solution and calcination time on sol-gel synthesis of hydroxyapatite. Journal of Bionic Engineering. 2019;**16**(2):311-318

[90] Farazin A, Ghasemi AH. Design, synthesis, and fabrication of chitosan/

hydroxyapatite composite scaffold for use as bone replacement tissue by sol–gel method. Journal of Inorganic and Organometallic Polymers and Materials. 2022;**32**(8):3067-3082

[91] Mandal T, Mishra BK, Garg A, Chaira D. Optimization of milling parameters for the mechanosynthesis of nanocrystalline hydroxyapatite. Powder Technology. 2014;**253**:650-656

[92] Bulina NV, Makarova SV, Baev SG, Matvienko AA, Gerasimov KB, Logutenko OA, et al. A study of thermal stability of hydroxyapatite. Minerals. 2021;**11**(12):1310

[93] Wang T, Dorner-Reisel A, Müller E. Thermogravimetric and thermokinetic investigation of the dehydroxylation of a hydroxyapatite powder. Journal of the European Ceramic Society. 2004;**24**(4):693-698

[94] Sofronia AM, Baies R, Anghel EM, Marinescu CA, Tanasescu S. Thermal and structural characterization of synthetic and natural nanocrystalline hydroxyapatite. Materials Science and Engineering: C. 2014;**43**:153-163

[95] Guo X, Xiao P. Effects of solvents on properties of nanocrystalline hydroxyapatite produced from hydrothermal process. Journal of the European Ceramic Society. 2006;**26**(15):3383-3391

[96] Djošić MS, Mišković-Stanković VB, Milonjić S, Kačarević-Popović ZM, Bibić N, Stojanović J. Electrochemical synthesis and characterization of hydroxyapatite powders. Materials Chemistry and Physics. 2008;**111**(1):137-142

[97] Manoj M, Subbiah R, Meena P, Mangalaraj D, Ponpandian N, Viswanathan C, et al. Green synthesis and characterization of bioceramic hydroxyapatite (HAp) nanosheets and its cellular study. Advanced Science, Engineering and Medicine. 2016;**8**(3):216-221

[98] Kim IS, Kumta PN. Sol–gel synthesis and characterization of nanostructured hydroxyapatite powder. Materials Science and Engineering: B. 2004;**111**(2-3):232-236

[99] Tõnsuaadu K, Gross KA, Plūduma L, Veiderma M. A review on the thermal stability of calcium apatites. Journal of Thermal Analysis and Calorimetry. 2012;**110**(2):647-659

[100] Gomes EDS, Lima AMD, Gavinho SR, MPF G, Devesa S, AAM M. Influence of polymeric blends on bioceramics of hydroxyapatite. Crystals. 2023;**13**(10):1429

[101] Kulpetchdara K, Limpichaipanit A, Randorn C, Rujijanagul G, Tunkasiri T, Chokethawai K. Microstructure-property relations of biphasic calcium phosphate obtained by hot pressing process. Processing and Application of Ceramics. 2019;**13**(3):300-309

[102] Sudarsanan K, Young RA. Significant precision in crystal structural details. Holly Springs Hydroxyapatite. Acta Crystallographica Section B. 1969;**25**(8):1534-1543

[103] Doebelin N, Kleeberg R. Profex: A graphical user interface for the Rietveld refinement program BGMN. Journal of Applied Crystallography. 2015;**48**:1573-1580

[104] Ben Taher Y, Moutia N, Oueslati A, Gargouri M. Electrical properties, conduction mechanism and modulus of diphosphate compounds. RSC Advances. 2016;**6**(46):39750-39757

[105] Lasia A. Electrochemical impedance spectroscopy and its applications. In: Modern Aspects of Electrochemistry. New York, USA: Kluwer Academic Publishers; 2005. pp. 143-248

[106] Macdonald JR, James R. Impedance Spectroscopy: Emphasizing Solid Materials and Systems. New York, USA: Wiley; 1987

[107] Devesa S, Amorim CO, Belo JH, Araújo JP, Teixeira SS, Graça MPF, et al. Comprehensive characterization of Bi1.34Fe0.66Nb1.34O6.35 ceramics: Structural, morphological, electrical, and magnetic properties. Magnetochemistry. 2024;**10**(10):79

[108] Gupta P, Mahapatra P, Choudhary RNP. Structural, dielectric, impedance, and modulus spectroscopy of $BaSnO_3$-modified $BiFeO_3$. Journal of Physics and Chemistry of Solids. 2020;**137**:109217

[109] Venkataraman BH, Varma KBR. Microstructural, dielectric, impedance and electric modulus studies on vanadium - Doped and pure strontium bismuth niobate ($SrBi_2Nb_2O_9$) ceramics. Journal of Materials Science: Materials in Electronics. 2005;**16**(6):335-344

[110] Kumar N, Patri SK, Choudhary RNP. Frequency-temperature response of a new multiferroic. Processing and Application of Ceramics. 2014;**8**:121-125

[111] Coşkun M, Polat AO, Coşkun FM, Durmuş Z, Caglar CM, Türüt A. The electrical modulus and other dielectric properties by the impedance spectroscopy of $LaCrO_3$ and $LaCr_{0.90}Ir_{0.10}O_3$ perovskites. RSC Advances. 2018;**8**(9):4634-4648

[112] Regadas JS, Gavinho SR, Teixeira SS, de Jesus JV, Pádua AS, Silva JC, et al. Influence of the particle size on the electrical, magnetic and biological properties of the bioglass® containing iron oxide. Magnetochemistry. 2023;**9**(9):209

[113] Jogiya BV, Jethava HO, Tank KP, Raviya VR, Joshi MJ. Impedance and Modulus Spectroscopic Study of Nano Hydroxyapatite. 2016. p. 020227

[114] Rehman F, Li JB, Zhang JS, Rizwan M, Niu C, Jin HB. Grains and grain boundaries contribution to dielectric relaxations and conduction of $Bi_5Ti_3FeO_{15}$ ceramics. Journal of Applied Physics. 2015;**118**(21):214101

[115] Faia PM, Jesus EL, Louro CS. TiO_2:WO_3 composite humidity sensors doped with ZnO and CuO investigated by impedance spectroscopy. Sensors and Actuators B: Chemical. 2014;**203**:340-348

[116] Bondarenko AS, Ragoisha GA. EIS Spectrum Analyser. In: Pomerantsev AL, editor. Progress in Chemometrics Research. New York: Nova Science; 2005

[117] Devesa S, Graça MP, Henry F, Costa LC. Dielectric properties of $FeNbO_4$ ceramics prepared by the sol-gel method. Solid State Sciences. 2016;**61**:44-50

[118] Ben Mohamed C, Karoui K, Jomni F, Guidara K, Ben RA. Electrical properties and conduction mechanism of $[C_2H_5NH_3]_2CuCl_4$ compound. Journal of Molecular Structure. 2015;**1082**:38-48